# Author's Note: The S

As I was turning off my computer and packing up my projector at the end of a Creation Seminar I recently presented, a young man approached me.

"I'm 20 years old," he said, "and I haven't been to church in three years but what you taught me tonight has completely changed my mind about Christianity. I wanted you to know that I will be back in church starting this Sunday."

He went on to explain that as he was growing up in his Christian church he became increasingly confused, frustrated and downright mad. The church had no answers to anything relevant in his life. He didn't think there were any viable answers to questions about science, evolution, old-earth beliefs, "there is no God," etc.

So this young man did what 81% of Christian-raised kids do these days—he stopped going to church. He also stopped believing God's Word.

When God called me to teach the Truth of God's Word and the science that supports the Biblical Creation account, I was disheartened and discouraged at every turn. The vast majority of churches would not even let me present the information.

The reasons I was given were many, but the bottom line was this: they did not believe in the Bible's accounts about our creation and they didn't care to hear why there was no reason to compromise on these issues.

They had become part of what I now refer to as the *Submerging Church*, a movement that is misleading hundreds of millions of Christians and seekers.

Today's Church is being sapped of its influence, vitality and sustainability. It is sinking, submerging like a leaky boat taking on water.

This report, *The Submerging Church*, reveals that to destroy true Christian faith, one must simply compromise or ignore the first few chapters of the Book of Genesis, especially the validity of a recent Creation.

We expose the depth to which compromises with secular opinions have embedded themselves into the *Submerging Church*, and we explore what can be done and how you can be part of the solution.

So…start turning the pages, keep an open mind and remember this truth:

**God always turns people TO His Word,
NEVER away from It.**

God bless.

*Russ Miller*

UCS PRESS PMB 119
1702 W. Camelback Rd. # 13
Phoenix, AZ 85015

UCS PRESS is an imprint of MarJim Books, Inc.

Bible verses and passages quoted in this report are
from the King James Version.

First edition, first printing, November 2009

Printed in the United States of America


ISBN: 978-0-943247-99-1

# Dedication

To all Believers who do not
compromise the Word of God.


# Acknowledgments

Special thanks to my
wife Joanna for her significant
input for this report; also to
Valeri Marsh, Jim Porter
and Brock Lee for their continued
insightful editing assistance.

# The Submerging Church

## Russ Miller
### with Jim Dobkins

## About the authors

After spending 18 years building a successful nationwide management recruitment firm, Russ Miller walked away from it all in the year 2000 so he could found Creation, Evolution & Science Ministries. He's dedicated his life to studying Creation-Evolution issues and developing CESM's popular, yet challenging, PowerPoint programs. Russ has presented over 1,000 seminars and church service messages and appeared on international telecasts. Through thousands of radio programs he has presented scientific evidences which have challenged hundreds of thousands of Christians worldwide to confidently believe the whole Bible – word for word and cover to cover. He lives with his wife Joanna in Flagstaff, Arizona.

Jim Dobkins is a Southwest-based writer. His co-author credits include *Winnie Ruth Judd: The Trunk Murders*, *Machine-Gun Man*, *The Ararat Conspiracy* and *My Boss was the BTK killer*. He conceptualized and wrote the dramatized short documentary *Someone Who Cares* and has ghostwritten several books. He has written over 1,000 articles. Primarily known as a crime writer, he now focuses on writing inspirational books and collaborating with Russ Miller in developing books in *The GENESIS Heritage Report Series*. This is the fourth report in the series, following *The Theft of America's Heritage . . . The Facts Are Talking, But Who's Listening?* and *371 Days That Scarred Our Planet*. He resides with his wife Marty in Arizona.

# Contents

9
ONE
We're Losing Them

22
TWO
The Church Has Become Irrelevant

43
THREE
The Fear Factor

64
FOUR
Jesus Christ:
The Ultimate *Non-Essential*

77
FIVE
Word for word,
cover to cover

91
SIX
Believers, Non-Believers
and Make-Believers

102
SEVEN
Hypocrisy Runs Rampant

# Contents continued

114
## EIGHT
## Relevance 101

129
## NINE
## Relevance 202

149
## TEN
## Courage

I was sitting behind my Creation, Evolution and Science Ministries resource table, loaded with books and DVDs that contain incredible scientific evidence supporting the Biblical, young-earth Creation and the Word of our Biblical God.

I saw him coming from 75 feet away. His fists were clenched, his face was red, and I thought *This guy is going to take a swing at me*. I stood and braced myself.

"What are you teaching?!" he demanded.

I looked him in the eye and said, "I teach that the Bible is true, word for word and cover to cover."

He backed off a bit, but continued, "I'm a college pastor and you can't tell college students that Carbon Dating doesn't work!"

"No," I replied, "You have to show them how it does work and how carbon dating proves the layers of the earth formed recently during the global flood."

Publisher's Note:  All pre-chapter quotes in this report are by Russ Miller unless otherwise cited.

# ONE

# We're Losing Them

**Consider these statistics.**

It is estimated at the time of this writing that in England, formerly a Christian-based nation, only 5% of the British attend a Christian church. It is predicted that in twenty-five years less than 1% of the British will be attending a Christian church.

This trend is continuing across the Atlantic.

In the United States of America, founded by predominantly Christian men and women on predominantly Christian principles, only 38% of Americans now attend a Christian church on a regular basis. That number is projected to shrink to only 15% of Americans attending a Christian church in 2035.

A generation of 20- to 30-year-olds has abandoned the Church. Studies reveal that 71% of children raised in Christian homes will leave the Church the day they move out of their parents' house. That count rises to 85% by the time they turn twenty years of age. (*Barna Research*)

A survey of 20- to 30-year-olds who regularly attended Sunday School as children revealed that a whopping 40% had their first doubts about God and His Word while attending middle school. By the time they graduated from high school that number had climbed to 90%. Yes . . . 90%. (*Beemer Research*)

**The most frightening statistic?**

The more *churched* a child is, the more likely he or she will be to leave the Church. It turns out that the very things we hope will keep kids in church, such as Sunday School, Vacation Bible School, church camps and retreats, are just not working. (*Beemer Research*)

**Why is the Church losing its children and down-trending so swiftly today, at a time when the world needs Jesus Christ more than ever?**

For all the upbeat music and praise bands, the small-group (*we've got your needs covered*) ministries, the state-of-the art buildings, sound systems, parking lots and coffee cafés, **we're still losing them.**

Add in the Sunday School hour, children's churches in finely-arrayed classrooms with teacher-friendly curriculum, over-the-top summer camps, exotic mission trips, one-on-one mentoring and fun-filled Vacation Bible Schools and, **we're still losing them.**

And let's not forget the multi-billion dollar *Christian industry* that offers music, DVDs, concerts, books, seminars, retreats, conferences, radio programs and TV shows.

*Yet we are still losing them.*

It doesn't seem to matter if you are a seeker-friendly, Bible-thumping, conservative or liberal-minded church. Your kids are walking out the door and the vast majority is not coming back.

In the book of Matthew there is a series of parables that focus on the sowing of the Word of God or, as Jesus puts it in Matthew 13:19, *the word of the kingdom.*

This is the first time parables are mentioned in the New Testament. Jesus gives us the Parable of the Sower in Matthew 13:1-9 and the Parable of the Wheat and Tares in Matthew 13:24-30. These are followed by the Parable of the Mustard Seed in Matthew 13:31-32 which is followed by the Parable of the Leaven in Matthew 13:33. Jesus provides detailed explanations of both the Parable of the Sower and the Parable of the Wheat and Tares to his disciples, which gives these two even further weight.

Why would Jesus begin his parable ministry with these particular examples? I believe it is because Jesus wants us to realize that the minute the Word of God is presented it will come under immediate and constant attack from the enemy.

I cannot emphasize enough the importance of these parables.

In fact I use the Parable of the Sower extensively in my live service messages and seminars.

> **Matthew 13:3-9**
> **3 Behold, a sower went forth to sow:**
> **4 And when he sowed, some seeds fell by the way side, and the**
> **fowls came and devoured them up:**
> **5 Some fell upon stony places, where they had not much earth: and forthwith they sprung up, because they had no deepness of earth:**
> **6 And when the sun was up, they were scorched: and because they had no root, they withered away.**
> **7 And some fell among thorns; and the thorns sprung up and choked them:**
> **8 But other fell into good ground, and brought forth fruit, some an hundredfold, some sixtyfold, and some thirtyfold.**
> **9 Who hath ears to hear, let him hear.**

However, for the purposes of this report, I will focus mainly on the Parable of the Wheat and Tares.

> **Matthew 13:24-28a**
> **24 Another parable put he forth unto them, saying, The kingdom of heaven is likened unto a man which sowed good seed in his field:**
> **25 But while men slept, his enemy came and sowed tares among the wheat, and went his way.**
> **26 But when the blade was sprung up, and brought forth fruit, then appeared the tares also.**
> **27 So the servants of the householder came and said unto him, Sir, didst not thou sow good seed in thy field? from whence then hath it tares?**
> **28a He said unto them, An enemy hath done this.**

> **Matthew 13:37-39**
> **37 He answered and said unto them, He that soweth the good seed is the Son of man;**
> **38 The field is the world; the good seed are the children of the kingdom; but the tares are the children of the wicked *one*;**
> **39 The enemy that sowed them is the devil; the harvest is the end of the world; and the reapers are the angels.**

In both the Parable of the Sower and the Parable of the Wheat and Tares, Jesus tells us of those who will come to destroy the planted Word of God. They are the birds (the fowl) and the tares, respectively. Both represent the minions of the evil one who undermine people's faith in the Word of God.

Notice that both the birds and the tares are found inside of God's earthly Kingdom. Apparently the tares are difficult to detect (but not impossible, as we read in the parable). Only the angels will be able to remove them. But we ARE told in God's Word that we can discern good from bad by the fruit (Matthew 12:33). And the fruit of the *Submerging Church* is bad indeed, as we will document throughout this report.

We will cover more fully the Parable of the Tares in the following chapters, but I would like to address one aspect of this parable right now.

> **Matthew 13:24-25**
> **24 Another parable put he forth unto them, saying, The kingdom of heaven is likened unto a man which sowed good seed in his field:**
> **25 *But while men slept,* his enemy came and sowed tares among the wheat, and went his way.**
> *(emphasis mine)*

# While men slept...

Over the past 200 years, while Bible-believing Christians were going about their earthly lives, others, whom I refer to as the tares, have been diligently working to destroy Christianity from within the Church itself.

While men slept, second by second the tares have been planting the seeds of doubt and compromise throughout the Church.

While men slept, minute by minute the tares have been recruiting members from within the Church to assist in leading Christians away from the foundational truths of Genesis.

While men slept, hour by hour the tares have been gaining ever-increasing influence among Christian leadership.

And while men slept, year by year the tares have been corrupting the doctrine of today's institutionalized Church.

But they didn't walk through the beautiful, stained-glass doors of churches with pitchforks in hand. No, they have been subtle, patient and amazingly efficient.

**They just got the Church to believe that the world was *billions of years old*. That's it. That's all it took and they have been hugely successful.**

I know what you're thinking... *That can't be it; I mean, come on, what's **a few billions years** got to do with it?*

Well, an old-earth accomplishes **two things** quite effectively:

**1]** It distorts the earthly evidence which supports God's Word;

   and

**2]** it deforms and devastates the Biblical worldview.

The evidence fully supports the Biblical worldview when the Biblical view is used to interpret the evidence. But while men slept, they failed to think clearly about the evidences supporting

Creation, the Global Flood, the Revelation of God throughout the entire universe and the Truth of Scripture. Befuddled and bewildered, we've slept through the *submerging* of the church and now we are paying a very high price.

## Awake thou that sleepest. (Ephesians 5:14)

I have found in my years of teaching that Christians, in general, operate on three levels:

> 1] They start out by holding a Biblical worldview so they readily see the evidence supporting a recent young-earth Creation; or

> 2] They start out by holding a secular worldview then they see the evidence for a recent young-earth Creation and adopt a Biblical worldview.

> 3] They start out by holding a secular worldview then they refuse to consider the evidence for a recent young-earth Creation and continue holding a secular worldview.

Each position is vitally relevant to the individual Christian and to the Christian Church as a whole.

**The primary goal of the tares is to take people away from worshipping the one true Jesus Christ.**

Tares don't care if they cause a person to reject Jesus or if they get a person to make up and worship a Christ not found in God's Word. Regardless, their most successful attacks focus on undermining a person's **Biblical worldview** or by distorting a person's understanding of the **evidence** which supports God's Word.

This is accomplished by bringing doubts about God's Word into the Church and filling the Church with several compromised forms of Christ as opposed to our one and only true Savior Who is presented to us in Scripture.

The tares are experts in the art of worldview warfare. They began today's ongoing assault by bringing into question the very foundations of God's Word which is a perfect Creation that was corrupted by original sin which separated man from God. (*Worldview Attack*)

They began framing their foundational attacks on Biblical Christianity at the start of the nineteenth century. This was when they inserted the concept that man's knowledge proved there was a need for *millions of years* beliefs in the place of God's six-day Creation. This concept, as we will see and have seen in other works of this series, has manifested itself in various versions of Christ, which we shall see below.

There was definite genius in their madness, as they did not go for the cross. Instead, they launched an all-out attack on the Biblical foundations established in the first few chapters of Genesis – the foundations upon which the Gospel of Jesus stands.

The weeds within the Church have reinforced their foundational attacks on the true Christian faith through unrelenting promotion of old-earth beliefs and to a certain degree, Darwinian philosophy, itself taking root in the latter half of the 1800's. (*Evidence Attack*)

The Church of the 1950's and 1960's did not have the benefit of today's abundant scientific evidence that thoroughly refutes Darwinism and old-earth beliefs. Many Church members wrongly assumed—and still assume today—that science was the unbiased search for the truth.

The tares used that lack of knowledge to help pour the foundational cement for atheistic Secular Humanism during 1962-1963, when teaching of Biblical Creation was removed from public schools and replaced with the religious philosophy of *Millions of years leading to Darwinian evolution.* (*Evidence Attack*)

Since then two more generations of unsuspecting children have been indoctrinated, mainly through public school textbooks, that the Biblical God is nothing more than a fairy tale. (*Worldview Attack*)

A current college textbook, *Discovering the Universe*, page 39, states:

> "The idea that the Earth was created in six days cannot be tested much less disproved. It is not a scientific theory but rather a matter of faith."

Can the idea that Planet Earth has been around for more than 4.6 billion years be tested any more than the idea of a six-day Creation?

Of course not.

The teaching of the philosophy of Darwinism in public schools, which went into overdrive beginning in 1963, led to the immediate shedding of God's moral constraints which He had designed to protect us from our own sinfulness.

By 1966 the sexual revolution and the drug culture took off. Violent crime soared. The radical women's liberation movement was birthed and grew emboldened and alternative lifestyle and radical animal rights movements took off. By 1973 abortion was legalized.

After all, people thought, if the scientific evidence proved Jesus did not create the world like He said He did, then how can the Scriptures be correct about my faith and behavior? If you can't trust the Bible about Jesus, what's the point?

I could write a book just on the evil fruits being produced by the notion that *millions of years led to Darwinism*. I cover many of them in my live teaching and DVD titled *The Evil Fruits of Darwinism.*

Old-earth beliefs, Darwinian teachings, a prosperous postwar America where a *What's in it for me?* attitude became contagious, and the disastrous *Don't discipline your children* philosophy of Dr. Benjamin Spock provided fertile soil in which America's children could be led away from the simplicity of trusting in the one true Jesus Christ.

Following World War II much of the Church began moving away from preaching on sin, the effects of sin and on judgment so people stopped seeing the theological conflicts caused by old-earth beliefs.

Not discussing sin has also led to churches filled with Christians who do not understand how vile our sin is to God. Without this knowledge we can not appreciate the depth of God's grace or the immensity of Jesus' sacrifice on the cross.

Charles Finney reportedly said:

> **"Evermore the Law must prepare the way for the Gospel. To overlook the instruction of souls is almost certain to result in false hope, the introduction of false Christian experience and to fill the church with false converts."**

Charles Spurgeon is quoted as saying:

> **"They will never accept grace until they tremble before a just and holy Law."**

Many of today's pastors, who only preach feel-good messages which are void of personal responsibility, provide fertile soil for the tares.

As these false concepts took root they led more and more Church leaders to unwittingly join the tares in compromising the Biblical foundations. Today, many Churches, Christian colleges and seminaries have built their house on the ever-shifting sands of secular opinions—instead of upon the Rock, the solid foundation of the Word of God. (*Worldview Attack*)

The resulting fracture of the Church into multiple beliefs on our origins has left Christians unable to muster support to defend Biblical truth or the United States of America's great Christian heritage. Today, Christianity's role in the founding of America has been virtually eliminated from public school history classes. We cover this in our report *The Theft of America's Heritage.*

The winds of change have allowed the *tares* to flourish within the Church and the born-again Christian community. These weeds have continually solidified their influence with ever-increasing numbers of people inside of the Church who compromise God's Word with anti-Scriptural, old-earth philosophies. Is it any wonder that we continue to say that such teachings undermine the authority of God's Word and corrupt the testimony of the Church?

Untold millions of true seekers have been led to stumble and fall due to the various old-earth heresies planted within the Church by the enemy himself. Stumbling blocks include Theistic Evolution, Framework Hypothesis, Revelatory Days, Day-Age, Gap Theory, Progressive Creationism and other old-earth-based notions that run contrary to the teachings found in God's inspired Word, the Holy Bible.

Over the past fifty years, while Secular Humanists were taking complete control of America's educational and scientific establishments, the tares have been equally busy obtaining advanced degrees and filling both accredited Christian colleges and seminaries with liberal theologians.

Within one generation, a small Southern Baptist university in the southwest deteriorated in its Christian outlook and was eventually sold to a commercial company. Though the school continues to advertise itself to be a Christian university, its teachings are filled with *millions of years* beliefs and many of its professors are non-Believers.

Many current—and most of tomorrow's—Church leaders hold to various compromised versions of Jesus Christ, none of which are found in the uncompromised Word of God.

Many churches have left the Christian *ministry* and are a part of the growing *Christian Industry.* Careers and salaries are the driving force of this growing industry and the Word of God is only used as window dressing to keep the dollars flowing. Everything is aimed at how many people attend and how much money is collected each week.

The leaders of the Christian Industry refuse to take on the Creation-evolution issues, keeping their flocks in the dark while refusing to shepherd their flocks against the wolf of this secular juggernaut. Almost to a soul they compromise God's Word with, or condone, the teaching of the secular foundation of *millions of years of time.*

Tainted beliefs are being passed on to their unsuspecting flocks in an ever-increasing downward spiral which is rearing its ugly head in the form of the *Submerging Church.*

A friend of mine's twenty-three-year-old daughter attended a church that brought in a world-renowned Progressive Creationist to speak. He planted the seeds of doubt about the age of the earth and of our origins in her mind.

At first she rejected the Biblical six-day Creation. Then others doubts sprouted forth as she began reading books from other old-earth believing sources. Within two years of that fateful Sunday service message she rejected her entire faith in Christianity. (And we tell good from bad by the fruit.)

Their great Christian heritage stolen from them, misled into thinking that the religious philosophy of *millions of years leading to Darwinism* is science, and seeing that the primary response of the Church has been to bow God's Word to man's fallible opinions, America's children have, not surprisingly, turned away from their Christian-based values.

As we pointed out early in this chapter, studies reveal that more than 70% of Christian-raised kids leave the Church the day they move out of their parents' house and by age 20 that number climbs to 85%.

False doctrine that is riddled with secular teachings comes with a very high price tag.

**Matthew 7:15 Beware of false prophets, which come to you in sheep's clothing, but inwardly they are ravening wolves.**

This report will expose the prominent position the Creation-evolution and age-of-the-earth issues have played in bringing about the many compromises with false Christs which are choking the faith from many people within the Church itself.

Though the majority of Christian churches, seminaries and colleges have already fallen victim to the tares, most of the victims fail to notice their corruption—so subtle has been the attack.

Yet only a remnant of Christian faithful stands weed-free today.

My prayer is that by shining the light of Biblical Truth into the dark shadows of the thorn bushes of Christian compromise, this report will alert Believers and true seekers to the dire situation within the *Submerging Church*.

I challenge Christians to review the foundational information discussed in this report, and allow God to use them in a most powerful way to help heal and reclaim the Church Body.

I'm standing in front of an expectant crowd of churchgoers during a Sunday morning service. I can see a few disgruntled looks, a few heads looking out the windows, not even wanting to look at me or hear what I'm about to say. It looks like a tough crowd. I know that some are thinking *I believe in an old earth* or *What does Creation and science have to do with church?*

I begin, "I have a few questions to ask all of you and please just answer these to yourselves. First, how many of you are worshipping a Theistic Evolution Christ who used evolution over *millions of years* of time to create the world? And how many of you are worshipping a Progressive Creation Christ that created the world *over millions of years* of time? Or how many of you are worshipping a Christ that used a day-age theory, or gap theory or framework hypothesis theory? Or, how many of you are worshipping a Christ who created the world about 6,000 years ago in a six-day Creation and judged man's sin with a global flood, just like the Bible says?"

A few people clap their hands. I've got everyone's attention now but some still don't look too happy.

I continue, "Now, I've got one more question for you and this one you can answer out loud. How many Jesuses died on that cross so your sin could be forgiven and you could be reunited with Him?"

Most of the disgruntled looks disappear as they answer in unison, "One."

I nod my head and respond, "Then I think we've got a problem in the Church that needs to be dealt with."

## TWO

# The Church has Become Irrelevant

You read this chapter title correctly.

It is tragic, but true.

It saddens and chills me to say this:

The Church has become irrelevant.

Now don't get me wrong.

God is not irrelevant.

Jesus is not irrelevant.

The Holy Spirit is not irrelevant.

Salvation is not irrelevant nor is the Bible irrelevant.

But the Church?

It has become irrelevant.

I will explain in this report why this is so and how it happened.

Over my years of ministry I have had many incredible volunteers offer to help in any way possible to support Creation, Evolution & Science Ministries (CESM). Many of these gracious people have tried calling churches, Christian camps and retreat centers to set up speaking engagements for me. Time and time again, the responses they receive are:

> **"Creation is a non-essential to our
> church family."**

or

> **"We believe in *millions of years* of time."**

And the topper:

> **"We had a Creation speaker here about six
> years ago, so we're covered."**

## The forgotten chapters of Genesis

A hallmark of the *Submerging Church* is omission of discussion of the first eleven chapters of Genesis.

After denying requests from church members to host my teachings, a staff member of the church presented a multi-week study of Genesis by omitting chapters one through eleven and BEGINNING with chapter twelve.

Why did he do that?

He had been so thoroughly brainwashed by the concept of *millions of years* beliefs that he had compromised God's Word with them. He obviously did not want the folks attending the study to know this.

And this is a church that was once one of the fastest growing and largest congregations in the United States. It had achieved that success the old-fashioned way via expository preaching from God's Word—word for word and cover to cover, including the Biblical foundations laid down in Genesis. Today the attendance at this church is only half of what it had formerly been.

## History of the universe and more

The origins of the universe, plant and animal life, and humans are all revealed very succinctly and powerfully in the first few chapters of Genesis.

We learn through the writings of Moses the *eyewitness* account of Creation by the Creator Himself.

Moses' use of the Hebrew language—which when studied in the context of how the words are used—clearly details Who created and how long it took our Creator to complete His creation process.

I urge you to read our report *371 Days That Scarred Our Planet* for a detailed discussion of the six-day Creation; how a global flood erodes all old-earth beliefs; and how science, including Carbon-14 testing, repeatedly confirms that God has recently judged man's sin with a worldwide flood.

The origins of sin, separation from God, and death also are revealed in Genesis, as is the model of marriage being between one man and one woman.

In those first eleven chapters of Genesis we learn about the results of Adam and Eve's disobedience in the Garden of Eden when they defiantly ate of the forbidden fruit of the Tree of Knowledge of Good and Evil.

That first act of willful sin by humans brought death into the world. It led to the devastation of Planet Earth via God's judgment of man's sin by the global flood that wiped out all human life with the exception of Noah, his wife, three sons and three daughters-in-law who lived in the ark for 371 days.

Again, I urge you to read our report *371 Days That Scarred Our Planet.* At the risk of being labeled biased, I believe that this report is the clearest, easiest-to-read and most understandable debunking of old-earth beliefs available. If you are sitting on the fence and are not really sure what you believe about how our universe was created, and how long it took God to create it, you will be able to use the abundant information provided and reach your own conclusion as to which side of the fence you are on.

### The foundation of the Gospel of Jesus Christ is laid down in Genesis.

Most Believers today can't tell you the precise verse in the Bible that first foretells the birth in human form of our Lord and Savior Jesus Christ and that He would be born of a virgin.

Well, here it is:

**Genesis 3:15 And I will put enmity between thee and the woman, and between thy seed and her seed; it shall bruise thy head, and thou shalt bruise his heel.**

The wording *her seed* (singular, not plural) tells us that the baby will be born of a virgin; otherwise, it would be the seed of man as is the case with the other mentions of *seed* in the Bible.

Really.

The foundation for the Gospel of Jesus Christ is laid out in the early chapters of Genesis. I call it the COSt.

## What is the COSt?

I will explain the COSt after the following brief discussion.

A question that most Christians are able to handle is *Why did Jesus Christ die on a cross at Calvary?*

The Biblical response is so our sins could be forgiven—the sinless blood of the Lamb of God permanently paid for the sins of all who truly believe in the Jesus Christ found in the Holy Word of God as their redeeming Savior.

However, a logical-thinking person might ask, "Well, if God made us and we are sinful, then why do we need to be forgiven for being the way that God made us?"

Well, in order to answer this from a Biblical perspective you must understand and be able to articulate the Biblical foundations which are laid out in the early chapters of the Book of Genesis.

I believe that God showed me a simple acronym to sum up the Biblical foundations given to us in Genesis. I refer to this as the **COSt** because the cost to redeem us with our loving Creator was the torturous death on the cross at Calvary that Jesus, God's only begotten Son, endured so that whoever believes in Him as their Lord and Savior will have everlasting life with Jesus in heaven.

So horrible was Jesus' suffering on the cross that it was excruciation – literally. The word excruciation comes from the Latin words *ex crux* – the cross.

## Here is the COSt:

**C** is for Biblical **C**reation as recorded in the Bible;

**O** is for the **O**riginal sin committed by Adam and Eve in the Garden of Eden;

**S** is for the **S**eparation from God that resulted from Adam and Eve's sin;

**†** in the shape of a cross stands for our **need of redemption** with our loving Creator.

**Now for the Biblical response to the last question:**

God didn't make us sinful.

In fact, He made us in His own image, rational creatures able to think and use our reason to understand the nature of things, beginning with who He was (our eternal Creator Who deserved our total trust and devotion) and who we are in relation to Him (His beloved Creation). He gave us a free will to either love Him or reject Him.

So when Adam and Eve were presented with two contradictory beliefs, "On the day you eat of it you will surely die" and the serpent's "You will not surely die," they should have used their God-given gift of reason to choose to believe the Creator rather than the created thing.

In doing so, they acted out of their rational nature, rejecting God, and bringing about a separation from Him which we all have inherited today. Of course, it also brought about the curse and all the accompanying suffering, which was now a necessary but merciful call-back to help us stop and think about our lives.

We have inherited this sinful nature from the first Adam and only the redeeming sacrifice of the last Adam, the Lord Jesus Christ, is able to reunite us, or redeem us, with our loving Creator.

God's perfect Creation having been corrupted by Adam's original sin, which separated mankind from our Creator, is the fundamental basis of our need for the Gospel of our redeeming Savior, Lord Jesus the Christ.

Destroy Biblical Creation and you will destroy the very reason for the Gospel.

It really is that simple.

Why else do you think Creation and the scientific evidences that support a recent, young-earth Creation are under unceasing, RELENTLESS attack (as are young-earth, Biblically-grounded Creationists)?

The fact is that the Christian Church has lost its foundation and is submerging rapidly. Though it is late, there is a vital difference between late and *too late*.

I believe the first true step toward a major reformation of the Christian Body would be returning to trusting in, and building the Church on, the never-changing rock of Biblical Creation. Then the Church would return to worshipping the one true Jesus Christ.

The great apostle Paul revealed the need for building on the Creation foundation when he spoke to the Greeks, as he reported in the books of 1 Corinthians and Acts.

**1 Corinthians 1:23 But we preach Christ crucified, unto the Jews a stumbling block, and unto the Greeks foolishness.**

In Acts 17:18 we learn that many of the Greeks thought Paul was some sort of crazed babbler after he had tried to explain to them how Jesus was killed and then resurrected in order to provide for their salvation.

Because the Greeks did not understand why they needed to be redeemed with their Creator they found Paul's teachings to be foolishness.

So the great apostle, never one to give up easily, met with them again on Mars Hill. As you read the following verses keep in mind that Mars Hill was named after Mars, who's Greek name is Ares, the god of war in the Greco-Roman religion. Paul was literally taking the fight to them, walking right into the domain of a war god and declaring the truth of the One True God. If only people had the faith and courage today to do a tenth as much.

**Acts 17:22-24**
**22 Then Paul stood in the midst of Mars' hill,**
**and said, Ye men of Athens, I perceive that in all**
**things ye are too superstitious.**
**23 For as I passed by, and beheld your devotions,**
**I found an altar with this inscription, TO THE**
**UNKNOWN GOD. Whom therefore ye**
**ignorantly worship, him declare I unto you.**
**24 God that made the world and all things**
**therein, seeing that he is Lord of heaven and**
**earth, dwelleth not in temples made with hands;**

Paul regrouped and began at the beginning by laying down the foundation of why the Greeks needed Jesus' redeeming sacrifice upon the cross. Paul went back to Creation. He related his teachings to the world the Greeks lived in.

Paul went back to the evidence, something the Greeks could touch, see and smell. He went back to the ground they walked on, the earth under their feet. He made God real, not a God of long ago or far away. He taught them of the God who created the heavens, earth and seas, and all that lives in them. And he no doubt told them how our sin had separated us from Him.

Only after the Greeks saw the relevance of the real God of Creation could they understand the true God of Redemption.

**Psalm 96:4-5**
**4 For the LORD *is* great, and greatly to be**
**praised: he *is* to be feared above all gods.**
**5 For all the gods of the nations *are* idols: but the**
**LORD made the heavens.**

A huge reason the Church is submerging into the mire of our evolutionary-based, secular society is that this vital teaching from the great apostle seems to have gone unnoticed.

Many of the Greeks still mocked Paul and rejected their Savior, as is always the case with rebellious mankind. Still, he was used by God to reap a bountiful harvest of saved souls and to plant the Christian Church in a society which was in desperate need of God's Truth and the love of the Lord Jesus.

It's sad to say that old-earth weeds are so deeply rooted throughout the Church today that it scarcely raises an eyebrow to announce you hold to non-Scriptural beliefs such as Progressive Creation, Gap Theories or Theistic Evolution. However, stand up and state that you believe in a world created during six literal days and judged about 4,400 years ago with a global flood (based on God's Word) and you may find yourself shunned, if not tossed out on your ear.

Think about it; why directly attack Jesus?

Some do, but with little success. After all, Jesus is a historical figure, well-documented in non-Biblical historical records.

But attacking the REASON for Jesus?

Now you're talking.

What a magnificently horrific plan, and this is exactly what Satan and his minions set about doing.

So, is their plan working?

Well, the proof of the pudding is in the eating and studies reveal tremendous amounts of bad fruit coming from the combination of Christian apathy and Humanist zeal which led to the Secular Humanists' triumph back in 1963.

Of American children born from 1927 to 1945 who graduated from high school before the teaching of Biblical Creation was replaced with the false science of *millions of years leading to Darwinism* in 1963, 65% believe the Bible is true.

Of America's kids born between 1946 and 1964 who graduated from high school as the first generation to be instructed exclusively in Darwinian teachings, just 35% believe the Bible is true.

Keep in mind that their teachers had been reared with the teachings of Biblical Creation.

**Then look at what happened.**

Kids born between 1965 and 1983 were the first generation to be taught Darwinism by teachers who, for the most part, had also been indoctrinated in Darwinian teachings. Only 16% graduated from high school while believing the Bible is true. (Statistics from TS Ranie; *The Bridger Generation: America's Second Largest Generation*; Broadman & Holman; 1997)

Young people are leaving their churches in droves because the Church has become powerless and irrelevant.

Church leaders know there's a problem. They see the numbers going down. They see the mass exodus of young people. But they appear unable, or unwilling, to grasp the foundational problem.

Once you understand why the COSt is foundational to the Gospel message you will clearly see why the anti-Christian crowd has had their crosshairs sighted in on Biblical Creation for the past 200 years—Biblical Creation is foundational to the very Gospel of the Lord Jesus.

**The Church doesn't understand the problem, nor do they want to understand it because they have accepted so much compromise on the issues and because, by golly, they've got the solutions.**

Let's look at some of the *solutions* the Church employs to cope with the problems of *irrelevancy*. These solutions cross all denominational lines, from conservative to liberal and even those non-denominational denominations.

I would also like to add a disclaimer at this time about the Church. I speak in all types of churches. I attend a great church in my town (when I'm in my town). I have been on church boards and served as Deacon, Trustee and Moderator. There are several pastors from various denominations on Creation, Evolution & Science Ministries' board. I love the true Church, but it's missing the boat on this which is largely why the Church is submerging. So please understand that it is out of deep concern for the Body of Christ that I write this report and lead CESM efforts to get the Truth to the troops.

## Solution Number One

### Our kids are leaving the church:
### we need a children's ministry pastor, a youth pastor
### and a college pastor.

I recall visiting several churches on the east coast. One church in particular had gone to a lot of time and expense to bring my wife and me to their church for a weekend visit, a mini-Creation blitz. Upon arriving, the pastor picked us up at the airport and on the way to the church he mentioned in passing (albeit, a bit embarrassed) that his youth pastor had picked this weekend to take the whole youth group on a skiing trip. They would not be attending any of our seminars.

This weekend had been on their calendar for months.

My best guess is that this youth pastor compromises God's Word with old-earth beliefs. Regardless of his reasoning, he left his youth to fend for themselves as they headed out into the secular world.

Sadly, this is not the exception with youth pastors.

I did a live radio interview on the highest-rated Christian radio station in Phoenix, Arizona early in 2009. The topic was why so many Christian kids are leaving the Church by the age of twenty, and several youth pastors were there. The radio host and I invited each of the leaders there—and the 170,000 listeners—to have me speak to their youth groups at any time – and for free (I never charge a speaking fee).

Not a single youth pastor responded.

This is the norm.

I always tell pastors that the children above the age of seven should be in the service to hear the information God has given me to share. Still, I have spoken in countless churches where the children remain in the worship service right up to the point where I begin speaking. Then they leave to go to their *scheduled classes*. Someone is not picking up on the relevancy of these issues.

Unfortunately, Sunday School classes rarely, if ever, teach them how the Word of God truly relates to the world they live in. And that world includes old-earth beliefs, Darwinian teachings, science and more.

These issues are taught to them in public schools and via the media through the worldview of Secular Humanism. As a whole, the Church sends the lambs to the slaughter by failing to prepare them to defend their faith.

**College pastors**

A small group of Christian students asked me if I would visit their campus and share my *50 Facts versus Darwinism in the Textbooks* teaching.

I rented an auditorium at my own expense (these college kids could not afford it) that held 450 people. I felt God would use me to reach many young Christians who were teetering on the brink of leaving their faith behind. Sadly, only one of the school's numerous college pastors agreed to send his students to the presentation.

Their excuses ranged from *We just don't have time to plan such an evening* to *I believe in evolution.*

At least the latter excuse was honest because when I show that the Secular Humanist foundation of *millions of years leading to Darwinism* is wrong it also reveals that compromising religious leaders are also wrong.

He didn't mind his group being misled, but he did mind them finding out that he was helping to mislead them.

When the evening of the event came around I shared the information with a full house of both college students and professors, many who came to defend the religion of Secular Humanism. I followed the teaching with an hour-long question-and-answer session.

During the Q&A time I responded to an anti-Christian statement by asking the crowd, "Why is it not acceptable to say such a rude thing about Islam or Hinduism or toward any other religion but it doesn't even raise an eyebrow to make such a statement about Christianity?"

You could have heard a pin drop.

I feared my answer to my own inquiry would be the last words I spoke on this planet. Still I proceeded, stating, "It is because Satan already has the followers of false religions, but Jesus Christ is the real deal so Satan and his minions hate true Christians."

I thought the place would erupt but silence filled the auditorium for several long moments before another question changed the subject.

The good seed had been deeply planted and many Christian students approached me afterward to thank me and to tell me how the information had helped strengthen their faith.

Following the event, which ran on the front page of the school's liberal paper for the next three weeks, several Christian students approached their campus pastors and asked them to invite me to visit and share our faith-building teachings with their group.

Not a single religious leader from the campus asked me to visit and help the young adults. Their excuses varied but centered on attacking me personally:

**"We cannot have him visit because he is mean-spirited...haughty...arrogant...unloving...not a scholar..."**

Such descriptions come directly out of the Postmodernism playbook in which we must tolerate all beliefs and act as if contradictory beliefs can all be true. Anyone claiming there is absolute Truth is labeled as mean, haughty, arrogant and unloving.

I refer to this as ABC tolerance:

**Anything But Christianity**.

I ran into two of these religious campus leaders shortly after the dust had settled and asked them why they said such horrible things about me. They responded, "Well, you came onto the campus and you told people who do not believe in Jesus that the Bible is true. That is very unloving."

Again, I point out that 85% or more of our kids leave the Church by the time they get out of college despite the fact we have more college pastors today than ever before in the history of the Church.

### Solution Number Two

**Our kids are leaving the church:
we need to entertain them.**

Come on, do we really want to try to compete with iPods, cell phones, guitar hero, and the Internet? Well the Church is giving it their best shot. Here are some current youth activities that I got off of church websites:

> Don't you hate it when church is boring? Well, so do we! That's why we have created a couple of great ways for high school students to connect here. Sunday mornings are a blast! We meet in the Student Center for a morning filled with laughs, great music, great friends, and relevant messages from God's word. The atmosphere is fun and inviting to make anyone and everyone feel welcome.

---

> For all 7th - 12th grade students. A week of crazy fun, intentional relationships and straightforward challenge for students who have completed grades 7-12. In the midst of ropes courses, worship services, paintball games, canoeing, recreation, and hanging out with friends, students will be presented with the truth about the "name" and the identity that God gives them.

---

We should always encourage the Body of Christ to fellowship with one another. But most youth do not need another entertainment outlet.

What they do need is to know that God's Word is true and they need to see the connection between fact and faith so that they see how the Bible and the Church are relevant to their lives. This is something only the Church can do.

One church website I visited had an article on how to find the faith *connections* in the TV show *Lost*.

And the relevance would be?

Today's Church has virtually thrown its children to the wolves. Deserted at the point of the secular attack on the authority of God's Word, and immersed in an educational system bent on teaching them what to think as opposed to how to think, Christian kids have been largely left to stand on their own against the philosophy of *millions of years leading to Darwinism*. (Teaching them how to think would make indoctrination very difficult.)

The fruit is that today only 4% of those graduating from high school in the United States of America believe the Bible is true.

But they do know how to rope climb, have paint ball wars and operate an iPod.

**Solution Number Three**

**Our kids are leaving the church:
we need better music.**

I grew up in the 1960's and 1970's. I know what music can do. Music can be a wonderful gift from God. It lifts our spirits. It calms our souls. But one thing music does not do is keep youth from leaving our churches. In a recent study only 1% of youth leaving churches left because of poor music. (*Beemer Research*)

Yet we have totally restructured our worship services to make music the focal point. We include drums, guitars, upbeat, lively music – mostly to show that we are culturally cool. In fact, our music does not sound much different from secular artists and bands. I have heard it said, somewhat accurately, that you just take a love song and replace the words *baby* or *darling* with *Jesus* in order to make it a Christian song.

I was speaking with a pastor whose church regularly hosted Christian musical groups. Though he said that several had been outstanding Christians, they were not going to host any more concerts as many were quite secular in the way they acted off stage.

He told me of one in particular who would only drink a certain brand of bottled water, only ride in a certain make and color of rental car and only stay in the finest hotel. He had had enough of their *star power*.

If all of this is to try to impress youth to stay in church I have a message for the Christian Body: It isn't working – kids are leaving the Church in huge numbers.

**Solution Number Four**

**Our kids are leaving the church:**
**we need to make the teachings shorter and sweeter.**

Today's sermons tend to focus on moral and spiritual issues, as many messages should do. However, they also tend to ignore the issues dealing with the world we live in – the earthly things that people can relate to such as our sinful nature and the proper apologetics which battle the foundational attacks and support our faith in the Word of God.

My Creation-based sermon messages are from 45 to 50 minutes in length. That is two to three times longer than the average message in today's *Submerging Church*.

I begin my teachings by setting the foundational issues laid out perfectly in the Book of Genesis. At the start, many eyes are wandering about the room, people are shifting in their chairs and the youth have a general look of disinterest.

When I begin to explain how the Book of Genesis has been corrupted by *millions of years leading to Darwinism* all eyes start to look my way. By the 10-minute mark, most folks, including the youth, are glued to the message.

This is because I'm talking about the anti-Biblical teachings that they get bombarded with on a daily basis and I'm refuting those false teachings. I'm relating God's Word to the world they live in.

Adults, youth and children are HUNGRY for this type of real-world Biblical teaching.

Studies show that 70% of all American adults WANT to believe in Biblical Creation. However, most of them have their doubts because they believe that the scientific evidences support an old earth and Darwinism.

When they see that the scientific data actually supports Creation, they begin to develop a Biblical worldview. This in turn will change the way they see the world, change the way they look at the Word of God, and improve their relationship with their one true Creator and redeeming Savior, the Lord Jesus Christ.

The Church desperately needs to rid itself of irrelevant, shallow, sentimental, *warm and fuzzy* teachings that have virtually nothing to do with people's daily lives. The Church needs to start connecting the Bible to the real world.

**Solution Number Five**

**Our kids are leaving the church: let's not call it the church anymore.**

Instead, let's call it community, or emerging streams, or café conversations. One idea behind *emerging* types of churches is to re-package it in a whole different look and language. Here's a quote on how the *emerging church* views the Word of God:

> **The Bible is not considered an accurate, absolute, authoritative, or authoritarian source**

**but a book to be experienced and one experience can be as valid as any other can. Experience, dialogue, feelings, and conversations are equated with Scripture while certitude, authority, and doctrine are to be eschewed! No doctrines are to be absolute and truth or doctrine must be considered only with personal experiences, traditions, historical perspectives, etc. The Bible is not an answer book.** (Brian, McLaren, *A New Kind of Christianity*, p. 52.)

This sounds eerily similar to Postmodernism mixed with just a pinch of Christianity tossed in for flavoring. I do agree with their assessment that the Bible is not just an answer book. It is much, much more. The Bible is the inspired Word of God. Todd Friel described Postmodernism as *simply making truth...an option.*

Remember:

God always leads you to His Word, never away from it. Believers need to compare what they are being taught or not taught—and their lives—to God's Word.

The *emerging church* is comparable to the Titanic; it's *submerging.* It's leaving true Christianity behind even faster than the rest of the Church.

### Let's not forget those Christian schools and seminaries.

I was excited about speaking to a large group of teachers and administrators from one of the world's largest organizations of Christian schools. Although I was speaking to a break-out group, the director had promised me that the next year I would definitely be one of the keynote speakers. Since this was a group of around 1,000 teachers, I was energized about the impact that would make. Even my break-out group was attended by well over 150 teachers.

About halfway through my teaching I noticed that one man got up and stalked out. At the break the director came up to me and said, "I'm sorry, Russ, but we can't have you be a keynote speaker next year. You'll upset too many teachers that believe in *millions of years.*"

I responded, "You mean, I'll upset the Christian teachers that compromise God's Word? Aren't they the ones who need to hear what I teach more than anyone?" He has never spoken with me since.

There are many Christian schools, colleges and seminaries that now teach old-earth philosophies while only a few teach their students how to reliably believe and defend the Biblical view of a recent Creation which has endured a global flood.

The fact is, the Institutionalized Church is churning out pastor after pastor who has been taught by their Christian professors and other *studied theologians* that *millions of years* of death brought man into the world. They then take this anti-Biblical teaching with them wherever they minister.

This is also a warning to parents. Don't get too comfortable sending your student to an accredited Christian college. There are only a handful of such schools that teach of the recent Creation which has endured a global judgment by water. More than 90% of them teach an assortment of old-earth beliefs which teach that it was not a loving God who created a perfect world, but, rather, it was a God who used long ages of death and suffering to bring mankind about. Of course these non-Scriptural teachings destroy the COSt.

Jesus had strong words for the Scribes and Pharisees who misled people through the teaching of man-made doctrine rather than the Word of God.

> **Matthew 15:7-14**
> **7 Ye hypocrites, well did Esaias prophesy of you, saying,**
> **8 This people draweth nigh unto me with their mouth, and honoureth me with their lips; but their heart is far from me.**
> **9 But in vain they do worship me, teaching for doctrines the commandments of men.**

10 And he called the multitude, and said unto them, Hear, and understand:
11 Not that which goeth into the mouth defileth a man; but that which cometh out of the mouth, this defileth a man.
12 Then came his disciples, and said unto him, Knowest thou that the Pharisees were offended, after they heard this saying?
13 But he answered and said, Every plant, which my heavenly Father hath not planted, shall be rooted up.
14 Let them alone: they be blind leaders of the blind. And if the blind lead the blind, both shall fall into the ditch.

———

Mark 7-6-9
6 He answered and said unto them, Well hath Esaias prophesied of you hypocrites, as it is written, This people honoureth me with their lips, but their heart is far from me.
7 Howbeit in vain do they worship me, teaching for doctrines the commandments of men.
8 For laying aside the commandment of God, ye hold the tradition of men, as the washing of pots and cups: and many other such like things ye do.
9 And he said unto them, Full well ye reject the commandment of God, that ye may keep your own tradition.

———

Although today they are not called Scribes and Pharisees their spirit lives on through the misleading teaching of compromising pastors, theologians, elders and other Church leaders. Much of this misleading teaching is centered on the omission of the Biblical foundations established in those first eleven chapters of Genesis.

In all honesty, I believe that many of today's compromising pastors and Church leaders do not realize the havoc they're wreaking on the Church Body. This is primarily due to their own brainwashing as they progressed through the public school system and then were exposed to further *millions of years* teaching in Christian colleges and seminaries they may have attended.

And many of these pastors and Church leaders do not get the brunt of critical and attack-mode comments that young-earth Creationists continually are subjected to. Here are two recent examples of such attacks I've received in e-mails to me through our website at www.CreationMinistries.org:

> **Hello Russ, you don't know it but it is the idiotic fundamental cases like you with pathetic mentally deficient brains that add to the earth's worst problems for humanity and God hates you. But Lucifer loves you for it.**
>
> ———

> **Russ, you sound like a true dolt, incapable of separating causality from superstition. Have fun making Americans more stupid and less capable of stopping global warming and loss of biodiversity which your invisible man doesn't seem to be doing a good job at.**
>
> **There is a time when grown up people need to give up their invisible friends and convincing other people god is real doesn't do any good, especially when you LIE about well established science. Your indoctrination of people who are emotionally weak is reprehensible. Still, have fun in the world of magical make-believe. Oh, stone any children lately?**

As I've always said, name calling is the last bastion for those who have no evidence. And apparently no intellect either.

**E-mail to Russ: Subject: AN UTTER EMBARRASSMENT**

As a Christian with a college degree I can't believe some of the claims you make in your videos and on your website. To see such things makes me doubt that you are even human because a chimpanzee could make better conclusions. There are flaws in evolution, but there are flaws in the Bible as well, ESPECIALLY with creationism. You embarrass Christians as you BLATANTLY SPEW GARBAGE FROM YOUR MOUTH and I can't help but be horribly enraged by the things you say. The Bible does NOT need to be taken literally and it's not right for you to tell people you're horribly under-educated opinion on things. It is a shame that people listen to you when you are a COMPLETE AND UTTER EMBAR-RASSMENT to Christianity. WOW.

**E-mail from Russ to Christian skeptic:**

Speaking of "horribly uneducated opinions," thanks for your e-mail. Your hate-filled message did not point out a single mistake, either from God's Word or from my teachings. I say you have none. If you know of one, please advise. In the meantime, try putting more faith in the Word of God and less faith in man's opinions. Russ

## THREE

# The Fear Factor

We have established that the Church has major *relevancy* issues. It seems to have a *self-esteem* issue to boot.

Since when did the Church become so thin-skinned?

Since when did the Church become so fearful?

Fearful of men, women, tithers, other pastors, parents, college professors, theologians, Secular Humanists, the media, political correctness, and, well, just about anything that moves – except for God and His Word.

In Psalm 27 King David told how he found his strength to stand against his enemies through his faith in the Lord.

> **1 The Lord is My light and My salvation; whom shall I fear?  The Lord is the strength of my life; in whom shall I be afraid?**
> **2 When the wicked, even mine enemies and my foes, came upon me to eat up my flesh, they stumbled and fell,**
> **3 Though a host should encamp against me, my heart shall not fear:  though war should rise against me, in this will I be confident.**
> **4 One thing I desire of the Lord, that will I seek after; that I may dwell in the house of the Lord all the days of my life, to behold the beauty of the Lord and to inquire in his temple.**
> **5 For in the time of trouble he shall hide me in his pavilion: in the secret of his tabernacle shall he hide me; he shall set me upon a rock.**
> **6 And now shall mine head be lifted up above mine enemies round about me:  therefore will I offer in his tabernacle sacrifices of joy; I will sing, yea, I will sing praises unto the Lord.**

The *Submerging Church* has indeed backed itself into the corner of irrelevancy by placing the ever-changing opinions of men, which are based on the secular worldview, over God's inspired Word. Fearing further condemnation from the secular world most of today's Church cowers at the mere possibility of being verbally insulted or slurred by them.

As a result the Body of Christ has largely lost its Biblical worldview and its backbone. Recent polls reveal that 95% of Christian adults do not hold a Biblical worldview. The same poll showed that 49% of pastors from *conservative* congregations no longer hold a Biblical worldview either. (*Barna Research*)

### Not conformed but transformed

Sure, Christians are called to turn the other cheek, love the sinner, not the sin, and...*walk with all lowliness and meekness, with long-suffering, forbearing one another in love*...(Ephesians 4:2)

However, we are also called to be a beacon of light and to not be conformed to this world but, rather, to be transformed by the renewing of our minds:

> **Romans 12:2 And be not conformed to this world: but be ye transformed by the renewing of your mind, that ye may prove what is that good, and acceptable, and perfect, will of God.**

A majority of the Church Body is omitting portions of the Word of God. This gives a distorted witness of Christ's character and nature, which is described in the Gospel of John 1:14:

> **And the Word was made flesh, and dwelt among us (and we beheld his glory, the glory as the only begotten of the Father,) full of grace and TRUTH. (emphasis mine)**

The *Submerging Church* spends much of its time being gracious, but refuses to defend, much less take up, the sword of TRUTH.

## WHY?

The Church has become timid and afraid. They worship at the altar of acceptance and approval. The *Submerging Church* is like a package of boneless chicken.

Theologian R.C. Sproul said it best:

> **"When I became a Christian, I understood that Jesus took my sin away. What I never heard from Him was that He intended to take my backbone away."**

We need to get our spines back.

The old saying that *Sticks and stones may break my bones but words will never hurt me* must have been penned at a time when people weren't so thin-skinned.

For at least the past fifty years name-calling has been the primary weapon of the secular crowd. This is due to two primary reasons: first, it is hugely effective; and second, as I previously stated, name-calling is the last bastion for those with no facts to support their position.

With the foundation of the secular worldview being *millions of years leading to Darwinism* Secular Humanists have to resort to name-calling.

Unfortunately, rather than learning to stand up for the Truth of the Biblical worldview, this has left the Church cowering from insults such as *You're so stupid* or other such linguistic marvels. This makes me wonder *what will happen when serious persecution comes to the Submerging Church?*

> **Revelation 2:10 Fear none of those things which thou shalt suffer: behold, the devil shall cast**

**some of you into prison, that ye may be tried; and ye shall have tribulation ten days; be thou faithful unto death, and I will give thee a crown of life.**

I believe hardcore persecution against Christians, prevalent in much of the world already, is coming to America soon—the type of tribulations that will test us by throwing many of us into prison.

And the Church can't stand up to name-calling?

When all it takes to get us to crawl back into our safe little non-confrontational church dens is for someone to call us *judgmental* or, heaven forbid, *unloving,* then the lion has truly been declawed.

The book of Proverbs addresses this issue:

**Proverbs 28:23 He that rebuketh a man, afterward shall find more favor than he that flattereth with the tongue.**

**Proverbs 27:17 Iron sharpens iron; so a man sharpeneth the countenance of his friend.**

Compromise cripples the Church and fear declaws Christians. Until we put an end to the compromises with secular philosophies within the *Submerging Church* by confronting them with truthfulness and conviction the Church will continue on its path to total irrelevancy.

### The Gospel according to Saint Ph.D.

For the record, I am not a Ph.D. I know many fine Christians that worked incredibly hard to obtain a Ph.D. and I think it is a fine accomplishment. Having said that, please do not think you cannot have an educated thought or idea about anything unless you have a Ph.D. diploma framed on your den wall.

**Acts 4:13 Now when they saw the BOLDNESS of Peter and John, and perceived that they were unlearned and ignorant men, they marveled; and they took knowledge of them, that they had been with Jesus.** (emphasis mine)

The Greek words literally say that Peter and John were *unlettered* men; just plain old *commoners*.

They were uneducated in the technical, rabbinical teachings. In other words, no Ph.D. for them! They were just plain, ordinary folks, fishermen and the like, for heaven's sake.

They didn't hold official positions nor were they escorted to sit at the head of the banquet tables. In fact, they were not even invited to attend the banquets. Yet they approached their callings with boldness and power because they had been called by God, and not ordained by man.

When we accept Jesus Christ as Savior and Lord we are with Jesus by the power of the Holy Spirit. We have the power to have understanding of God's Word and properly understand the scientific evidences that support the recent Biblical Creation as the only viable option for our origins. We then—again through the power of the Holy Spirit— have the power to be bold and proclaim that information to any and all that may question it.

Jesus Christ was never ordained by man. The much-studied religious leaders and scholars of that time hated Him for it.

**Matthew 21:23 And when he was come into the temple, the chief priests and the elders of the people came unto him as he was teaching, and said, By what authority doest thou these things? and who gave thee this authority?**

The same spirit is alive today.

The *Submerging Church* has lost sight of this Biblical truth, often taking the teachings of secular man over the well-supported Word of God.

I have poured over a decade of my life into the study of science (biology, geology, paleontology, archeology, astronomy, and more) that supports the Biblical, young-earth Creation.

I have read hundreds of books, given thousands of seminars, spoken on colleges campuses holding open Q&A sessions, answered innumerable e-mails, written four books, held live call-in radio programs and appeared on international televisions programs yet skeptics within the Church often undermine my efforts by asking others *By what authority does he teach?*

When I felt the call of God on my life to teach the Church Body how real science supports the foundational Truth of Scripture, I set a goal to make my teachings enjoyable and understandable. I wanted to communicate information at a popular level that almost anyone could understand and follow. I believe this is what God gifted me with and called me to do.

Still, some people do not appreciate the simplicity in how I present the facts:

> **Mr. Miller, I am a college educated Christian and a firm believer that creationism is nonsense. I reconcile this with my religion in that God keeps time very differently than me. Please face the facts that man evolved over millions of years from lower forms of life and that there never was a Garden of Eden nor an Adam and Eve. To follow such ideas is to delude yourself.**

I can imagine what you are likely thinking but please note one important thing: this fellow is a victim of the *Submerging Church* which has failed to stand firm for the uncompromised Word of God.

Many of today's Christians have been led to think they can simply believe, or not believe, whatever they choose. They expect a buffet-style religion—a little of this, a little of that, mix and match as you please. This is simply open rebellion against God and His Word and it has been fostered by the spineless apathy of the *Submerging Church.*

I truly believe one goal of the Humanistic-dominated science community is to make science difficult to understand. Then, through the haze of confusion, they make it sound as if only those with years of experience in genetic research, who also have access to an electron microscope, can see the evidence supporting the religious belief in *millions of years leading to Darwinism.* They then work to intimidate others, including Christians, into thinking that we must simply humble ourselves to *their* word.

### And just Who are YOU to question THEM?!

Well you, my friend, have the greatest, most intelligent Teacher in the entire universe:

Jesus Christ.

The Creator Himself has given you His true history Book (the Bible) about His Creation. Christ has also given you the mind to understand His Word and the Holy Spirit to reveal all Truth.

> **2 Timothy 1:7 For God hath not given us the spirit of fear; but of power, and of love, and of a sound mind.**

Empowered with knowledge of the Truth of Jesus Christ, and by the presence of His Holy Spirit, the Church needs to rebuke the spirit of fear and reclaim the spirit of power, love and of a SOUND MIND.

> **2 Timothy 3:16 All scripture is given by inspiration of God, and is profitable for doctrine, for reproof, for correction, for instruction in righteousness:**

But frightened into compromising God's Word due to the false science of the secular crowd, and lacking a true and reasonable faith in God's Word, the *Submerging Church* is missing the lifeboat. A sound mind that can understand complex scientific data is a gift from God, as long as we get rid of that debilitating spirit of fear.

Further, anyone with a sound mind can reason just from the Creation alone, without any scientific data at all, that God must have created the universe. Think about it clearly for long enough and you will conclude that things don't just pop into existence from out of nowhere. There must have been an eternal creator. This itself flies in the face of godless Darwinism.

One sure sign of the *Submerging Church* is that it will allow its people to follow the broad path of secular teachings in the hope of not upsetting anyone.

As long as the Church fears rocking the sinking boat, questioning the professor, or believing the Bible is completely true, and, as long as the Church cowers in its church buildings, it will never get the spirit of power or a SOUND MIND.

It will continue to sink into irrelevancy and oblivion.

Scripture has much to say about this.

> **Colossians 2:8 Beware lest any man spoil you through philosophy and vain deceit, after the tradition of men, after the rudiments of the world and not after Christ.**

God knew this would be a huge problem.

### Philosophy and vain deceit

In the movie *Expelled: No Intelligence Allowed* Ben Stein took a look at how the secular establishment rages against honest researchers and teachers within both the scientific and educational establishments. The battle cry of the secularists who dominate these institutions is:

**Toe the *millions of years leading to Darwinism* line or be expelled.**

Failure to toe that line has resulted in lost careers, lost wages, missed promotions, denial of tenure, denial of publication, failure to receive grant money to continue scientific research, and a host of other punishments.

Researchers are not free to take their findings to their logical conclusion when they conflict with secular beliefs, much less when discoveries point to our Intelligent Biblical Designer. Thus the name of the documentary: *Expelled: No Intelligence Allowed.*

As I chronicle in report number two, *The Facts Are Talking. But Who's Listening?*, nothing short of lawsuits, coercion and blackmail are keeping the Darwinian ship afloat. Then, as I show in report number three, *371 Days That Scarred Our Planet*, old-earth beliefs are religious in nature as opposed to being a provable fact.

Yet because of the lack of knowledge of this information the *Submerging Church* is as frightened of the secular scientific community as are the blackmailed educators and researchers.

The secular battle cry aimed toward the *Submerging Church* seems to be:

Toe *millions of years leading to Darwinism* or we will insult and belittle you.

And why not; it works!

The last time I spoke at a university, presenting *Science Versus Darwinism*, an astrobiology professor reacted by starting an anti-Russ Miller blog site followed by an anti-Creation course. This is an actual accredited class titled *Science and Creationism.*

Two people I know took the course and reported that the class was not, as had been advertised, an open forum to discuss the science that supports Creation. No, not hardly. It was a class that berated and insulted Christians for believing in Creation. In fact, for their final exam they simply belittled me for 90 minutes.

All I can say is, "So what?"

I will say it again: name-calling is the last bastion for those with no real evidence to back up their position.

In my opinion this professor had too much time on his hands.

Think about the situation.

The Secular Humanist community owns the textbooks, the public schools, the universities, the scientific establishment, the museums, the national parks, the media and quite a bit of the *Submerging Church.*

Why would my one-hour teaching on a secular college campus warrant an accredited class attacking Biblical Creation?

Is their position really that weak and fragile?

Yes, it is.

Yet because the Church has failed to learn the science that refutes the secular position while largely refusing to stand up for God's uncompromised Word, the Church has become irrelevant.

Too busy putting on concerts and cowering in the corner, the *Submerging Church* has become irrelevant in people's daily lives.

One well-known Christian magazine published an article in its September 2004 edition calling for Christians to:

**Stop attacking evolution; start supporting the Intelligent Design Movement (IDM) and begin attacks against young-earth Creationists.**

This *Christian* publication called for people to support both the secular, anti-Biblical interpretations of the world we live in and the non-Christian IDM. Then it actually called for professing Christians to begin attacking Bible Believers. Fear has gripped the minds of the editors of this publication.

**Mind-boggling!**

**Timothy 6:20-21**
**20 O Timothy, keep that which is committed to thy trust, avoiding profane and vain babblings, and oppositions of science falsely so called:**

**21 Which some professing have erred concerning the faith.**

I often point out that real science is a Believer's best friend, as Scripture is continually being supported by new finds and has never been refuted by scientific data. The evidence and research lines up with the Word of God perfectly because the Word of God is TRUE, word for word and cover to cover. For a professing Christian to believe otherwise is certainly an example of someone erring concerning their faith.

## Fear on Campus

So what about the 90% of accredited Christian colleges and seminaries that have kowtowed to *millions of years beliefs*, teaching future Church leaders to trust in various old-earth philosophies which undermine the COS✝?

Well, they are doing something far worse than simply erring in their own faith. They are causing their students to stumble onto the broad path that has led to today's irrelevant, *Submerging Church*.

> **Matthew 18:6 But whoso shall offend one of these little ones which believe in me, it were better for him that a millstone were hanged about his neck, and that he were drowned in the depth of the sea.**

Even worse, these institutions are providing false doctrine which will be spread throughout the Church as these students graduate and go into the ministry, unintentionally misleading their flocks.

> **Luke 6:39 And he spake a parable unto them, Can the blind lead the blind? shall they not both fall into the ditch?**

Unless the situation is reversed soon the harvest will be the end of the true Christian faith on Planet Earth.

We have established that the *Submerging Church* is fearful and cowering because it wants to avoid more disapproval from the secular world.

We have also established that this has resulted in the Church becoming irrelevant by failing to defend the Word of God, and even worse, by actually joining in and teaching anti-Biblical doctrines.

The results of solid research reveal that the Church is completely failing to stem the mass exodus of its younger generations.

There is one last issue that is important to cover as we explore the reason the Church is submerging.

**It's a matter of pride.**

**E-mail: Pastor to Russ:**

**Russ, we need to be careful about being overly simplistic in our interpretations of Scripture on any scientific matter. While I'm no geologist, looking at glaciers along with violent up thrusts of entire mountains, I definitely get the impression it took more than a few thousand years to happen. Where does an ice age or two fit into a 6,000-year-old earth? It sure looks to me like a longer history. Does the Bible require us to take a 6,000-year-old interpretation? I would argue that we are not limiting God to create by His Word if we say that it seems the earth is old.**

Note how he believes that his personal opinions trump the Word of God. This kind of mindset is another common error within the *Submerging Church*.

Note also that, initially, this pastor had told me he believed in an old universe with a recent creation of the earth. Now he was claiming that the earth itself was quite old. I had already covered these issues in a multiple-page e-mail and in our DVD teachings which he did not look at.

**Russ to pastor:**

**When I look at the canyons, mountains, volcanoes and strata layers, I see a young earth that God judged with global flood. I don't know of any scientific reason to believe in an old earth (discounting the inaccurate isotope dating methods and the non-existent geologic column). I would be glad to sit down and discuss my thoughts about the earth's surface with you should you wish. God bless, Russ**

The pastor then forwarded my comments to a church elder who is a Gap Theorist. This is a person who believes that between Genesis 1:1 and 1:2 there is a *Gap* in which God made a different creation than the one found in the Bible, destroyed it because of Satan and his minions and then made the creation found in Scripture…while leaving Satan and his minions in it…and called it very good…

**Gap Theorist to Russ:**

**Russ, Pastor forwarded your e-mail to me. I would offer the following comments:**

**The age of the universe (scientifically) is based upon an astrophysical application of Einstein's general relativity and the "Big Bang" and is viewed as about 15 billion years old. The 4.6 billion year age of the earth is based upon**

**astrophysical calculations and over 40 radiometric dating techniques.**

Did I mention that I truly believe one goal of the Humanistic-dominated science community is to make science difficult to understand? His assertions are total hogwash and are based upon his faith in the word of researchers whose interpretations are based upon the secular worldview.

**Russ to Gap Theorist:**

**It seems that the discussion has gone from the church leaders believing in an old universe with a recent Creation, to believing in an old earth. The latter is a huge issue for Biblical accuracy.**

**Modern old-earth dating is based on the denial of the global flood as prophesied in 2 Peter 3:5-6.**

**This is due to the belief that the sedimentary layers, which make up the crust of the earth, formed slowly rather than as a result of the global flood judgment of man's sin. Since these layers are full of dead things (flood victims), such old-earth beliefs place death before man's sin. This in turn corrupts the accounts of original sin, the resulting separation from our Creator, the entering of death into God's perfect Creation, and why Jesus needed to die on the cross (to redeem us with our Creator).**

**The radiometric techniques are based on multiple wild guesses. If any of the guesses are wrong, and the odds are that all are in error, it will throw the results off by millions or billions of years. For instance, they assume the daughter**

material being measured was not present when the rock first formed and that the rate of decay has always been the same. Each wild guess is highly suspect.

Published dates fit the same range because they "select" them from the geologic time scale which is based on their belief of gradual strata formation. Dates that don't fit their time scale are ignored.

With regard to Carbon Dating, most scientists agree C-14 should be gone in measurable amounts in 70,000 years. Yet recent studies show that all fossil-bearing layers (580 million years old per the geologic time scale) still contain C-14 and in the same range of amounts from the top to the bottom layers! This proves the layers are only a few thousand years old and formed during the same event (the global flood). This in turn destroys the geologic time scale and all old-earth beliefs based upon it.

With regard to using the Big Bang to calculate the age of the universe, we are now on our fourth Big Bang of the past 80 years. Each has been taught as being factual long after being scientifically refuted, as is the current one. A letter signed by dozens of scientists appeared in *New Scientist* (Bucking the Big Bang, May 22, 2004). It included statements such as: The big bang theory can boast no predictions that have been validated by observation. Claimed successes consist of retrospectively making observations fit by adding adjustable parameters. The big bang relies on a growing number

**of never observed entities...and can't survive without these fudge factors...In no other field of physics would this continual recourse to new hypothetical factors be accepted. (Colossians 2:8).**

**Everyone has the same evidence to look at but our beliefs bias how we interpret the facts. I show people that God's Word fits the scientific facts perfectly when the evidences are interpreted through the Biblical worldview.**

**I cover the overwhelming evidences of the global flood, including stalactites/stalagmites, strata layers, continental drift, mountain ranges, geologic compression events, the *Three-Day Formation of Grand Canyon*, the ice caps and more much more in our DVDs and website. God bless, Russ**

This detailed, Biblically-based and scientifically sound explanation did not budge the pastor or the gap theorist from their non-Scriptural positions.

**Neither man ever referenced a single Bible verse.**

Instead they sent me an e-mail calling me an ungodly man (seven times), uneducated and other derogatory names – including *unloving*.

**Proverbs 9:8 Reprove not a scorner, lest he hate thee: rebuke a wise man, and he will love thee.**

Operating in ignorance may be one thing, although certainly not a good thing. However, to continue on trusting in or teaching a false doctrine after being shown it is Biblically incorrect is quite another issue.

Why would a pastor do this?

Why would a youth pastor refuse to allow the kids he is supposed to be shepherding see our seminars?

Why would any professing Christian block young-earth, Biblical Creationists from sharing the scientific evidence that not only supports the undeniable Truth of Scripture, but makes God's Word totally relevant to the world we live in?

**Pride. Pure and simple.**

They have been taught anti-Biblical, old-earth teachings and refuse to consider that they may have been misled. Thus, no matter how much Scripture or scientific evidence can be presented that refutes what they have been taught, or supports exactly what the Bible says, or relates God's Word to the world we live in, they choose their vain pride over the Truth of God's Word.

They are too proud of their *education,* of their *knowledge,* to admit they have been deceived into doubting God's Word which has always been correct. Sadly, they lead the sheep that trust them onto the broad path that leads to destruction.

> **1 Peter 5:5b-6**
> **5:b God resisteth the proud, and giveth grace to the humble.**
> **6 Humble yourselves therefore under the mighty hand of God that he may exalt you in due time.**

This Biblical principle of God giving grace to those who humble themselves to His Word is also found in Proverbs 3:34 and James 4:6.

We have established that the Darwinian scientific community has a stranglehold on all public—and to a certain degree Christian—education.

*Millions of years* is so engrained into the *DNA* of modern education that most people do not even question where these beliefs are derived from. I mean, come on, we are too smart to be misled today! We seem to think that no matter what the Bible says, if *millions of years* were not so, if lies were in the textbooks, we would have been able to pick up on that. There is no way that false and misleading teachings could have gotten past our brains.

Right?

### Pride goeth before a fall.

Anyone in any scientific field has been indoctrinated with old-earth teachings, including professing Christians who are scientists. Even the vast majority of pastors who have attended a seminary have been taught *millions of years* philosophies.

For someone with this amount of education to admit that they could have been fooled by such fraudulent teachings would take quite a bit of humility.

To admit that we believed a lie and to come to repentance of that error requires that we eat a huge chunk of humble pie. And the higher the level of education is the bigger the chunk.

Sadly, few leaders seem to come to that place of humble acceptance of having been fooled.

The result is that millions of Christian-raised kids will leave the Church. Many will lose their testimony entirely. All because of prideful Christian leaders who place their egos over the Truth of God's Word, making the *Submerging Church* irrelevant in the process.

The *Submerging Church* needs a heaping helping of humble pie. However, the overall response of the Church has been to either try to ignore what is being taught every single day to their flocks, or even worse, to try and blend such secular, anti-Biblical teachings into God's inspired Word.

The *Submerging Church* has placed a virtual blackout over teaching Christians to stand up for their faith, and this spineless cowardice has led the past two generations to view the Church as being non-relevant to their daily lives.

Kids are smart.

They realize that a person does not compromise what they know to be true.

They know *millions of years* conflicts with Scripture. They realize their church has failed to show them why God's Word is true and why the secular interpretations of the world are off base. And many of them are mad at the Church for failing them.

I lead Biblically-based Grand Canyon Bus Tours. While purchasing tickets at the Grand Canyon IMAX for one of our groups the woman selling the tickets asked me, "What group are you with?"

When I responded, "Creation Ministries.org" the twenty-year-old women next to her jumped in with, "And just what are you teaching those people because science has refuted everything I was ever taught in church!"

Was she angry?

No, she was irate.

And she had every reason to be angry at the *Submerging Church* which had failed her, and millions of kids like her, miserably.

Years ago I was a Theistic Evolutionist. That's a person who believes that God used evolution and *millions of years* to get us here.

When someone cared enough about me to tell me that I was misled I thought, *Yeah, right! I have 170 college credits. I began and own my own nationwide business. I'm a trustee of my church. What do you mean I was misled?*

However, I was willing to look at the evidence and the scales quickly fell off my eyes.

My life was changed forever as I humbled myself to the magnificent and TRUE Word of God. My faith skyrocketed.

As I began to study the scientific evidence I always compared it to the Word of God. I used God's Word as the filter, the looking-glass.

What I found was this:

What we learn about the physical world that we live in through real science meshes perfectly with what we read in Scripture, word for word and cover to cover.

Yet when we want our children to learn about the physical world they live in most Christians send their kids to a school that teaches them to see the world through the secular worldview. As we hurry about earning money our kids are being taught that life evolved without God over *millions of years* of death and that there was most definitely never a global flood.

Meanwhile, the Christian Church has failed miserably to relate God's Word to the world we live in. This is why the great majority of Christian-raised kids are leaving the church by the age of twenty.

This is why the Church must stop holding onto the traditions of men and change the way they teach if they truly care about shepherding the flock. And if this change does not occur very soon, it will be too late.

As I stated, I was formerly a Theistic Evolutionist. I ate my piece of humble pie and trust me when I say that the results of doing so are delicious.

One of my first speaking engagements was a series of Creation seminars at a large, seeker-friendly church. I only got invited to speak because several leaders of the church, excited about what I have to share, set up the event.

It turned out that the senior pastor was not happy about this and during a pre-seminar meeting he warned me, "I'm okay with most of the Creation evidence, but I do not want you to talk about the age of the earth."

I looked at him and replied, "I don't think I can do that. Old-earth issues are the main cause behind so many people losing their faith in the authority of God's Word."

He looked around the room at his staff members, several of whom seemed perturbed I was even breathing, and replied, "We believe that there are some elements of Scripture that are vitally important. However, there other areas that are non-essential and aren't worth debating. We consider the age of the earth to be a non-essential."

Later I learned that one of the staffers was a Theistic Evolutionist and another was a Gap Theorist. I never did ascertain what the senior pastor believed.

On the first evening I presented *50 Facts Versus Darwinism* to 350 people and the energy level was through the roof. However, as I ended my closing prayer the senior pastor walked up from behind me, took the microphone and said, "People, people...settle down, settle down. You don't need to believe anything this guy said."

The energy left the room as if a pin had pierced a balloon. Apparently, to this pastor, Biblical Creation was another *non-essential* part of Scripture.

# FOUR

# Jesus Christ:
# The Ultimate *Non-Essential*

Two-thousand years ago many people followed Jesus seeking to be healed. They wanted healing rather than the Messiah.

Today many people go to church wanting salvation but they really do not want Jesus. As a result, just who Jesus truly is has become a *non-essential* in the *Submerging Church*.

The fact is that whether or not Jesus is who He claims to be and has done what He claims He did is *most essential*.

## Jesus as Creator

> **Colossians 1:16-17**
> **16 For by him were all things created, that are in heaven, and that are in earth, visible and invisible, whether they be thrones, or dominions, or principalities, or powers: all things were created by him, and for him:**
> **17 And he is before all things, and by him all things consist.**

As we've cited, studies have established that by the age of twenty, 85% of America's Christian children will no longer attend church. This number is rising and this reality alone is strangling the *Submerging Church*.

We have documented how the Church has made itself irrelevant through compromise and cowardice. It is sinking fast and dramatic changes are needed. More of the same old milk toast sermons, tried-and-failed Sunday School programs, hand-clapping worship songs or rock star praise bands are not going to resuscitate the drowning crew.

Unfortunately, the Church has become too fearful to make the necessary changes. Christians are afraid of the slings and arrows of insults, and of the bullying from the secular world. Church leaders are afraid of offending someone who is tithing to *their* church. After all, they are barely making ends meet as it is, for goodness sake!

The result is the *Submerging Church* remains slunk back into its seemingly safe little corner pretending that all is well.

But all is not well.

Not when the majority of our kids are leaving the Church behind. Not when Christian principles are disappearing from our society and around the world.

The *Let's not rock the boat* attitude, fueled by the desire to be acceptable in our secular world, has led to a level of cowardice and compromise which has made the Church irrelevant. The pretentious pride and self-importance of misled leaders has pushed the *Submerging Church* into even deeper and more dangerous waters. The situation is at a critical stage yet hardly anyone seems to be awake...*while men slept.*

As I travel the United States of America, I am constantly befuddled as to why Church leaders will not embrace the scientific proofs supporting both Creation and the global flood as espoused in Scripture so succinctly.

John 3:16 is THE banner verse for Christianity. If you wanted to sum up the whole Gospel message in one verse, most folks would take you to John 3:16:

> **For God so loved the world, that he gave his only begotten Son, that whosoever believeth in him should not perish, but have everlasting life.**

**Whosoever <u>believeth</u> in HIM**

We must believe in Jesus Christ as our Lord and Savior to be saved from our sins and from spending eternity in hell.

But *Who* is this saving Jesus?

Throughout Scripture Jesus is given many names, including:
Alpha and Omega;
Bread of Life;
Chief Cornerstone;
Christ;
High Priest;
Immanuel;
King of Kings and Lord or Lords;
Lamb of God;
Light of the World;
Lord;
Mediator;
Messiah;
Prophet;
Rabbi/Teacher;
Savior;
Shepherd;
Son of David;
Son of God;
the Word;
and CREATOR.

**Jesus Christ as Creator**

> **Psalms 148:5 Let them praise the name of the LORD: for he commanded, and they were created.**

_____

> **Isaiah 43:1 But now thus saith the LORD that created thee, O Jacob, and he that formed thee, O Israel, Fear not: for I have redeemed thee, I have called thee by thy name; thou art mine.**

_____

**John 1:1-3**
**1 In the beginning was the Word, and the Word was with God, and the Word was God.**
**2 The same was in the beginning with God.**
**3 All things were made by him; and without him was not any thing made that was made.**

————

**Ephesians 3:9 And to make all men see what is the fellowship of the mystery, which from the beginning of the world hath been hid in God, who created all things by Jesus Christ:**

————

**Colossians 1:16 For by him were all things created, that are in heaven, and that are in earth, visible and invisible, whether they be thrones, or dominions, or principalities, or powers: all things were created by him, and for him:**

————

**Revelation 4:11 Thou art worthy, O Lord, to receive glory and honour and power: for thou hast created all things, and for thy pleasure they are and were created.**

There is no doubt in Scripture that Jesus is the Creator. Each one of these Scripture verses gives a significant amount of information about Jesus as Creator. I cannot over-emphasize the importance of believing in the only true Jesus Christ as described in the Word of God because THIS is the only Jesus who died on a cross in order to redeem us with Him.

Believing in the one true Jesus is what I call *essential*.

A quick look in Strong's Concordance teaches that the Greek word *pisteuo* which is translated as believe or believeth. It means to *put one's trust in*, with the implication that actions based on that trust will follow.  If we believe in Jesus Christ as our Creator, Lord and Savior certain actions based on our faith will follow.

In Webster's 1828 dictionary the word *believe* is the combination of two words: *be* and *lieve*.

*Be* is a prefix meaning: *on, around, over, thoroughly* and also, *make or cause to be.*

The word *lieve* is a word closely related to the word *live*. Among the many meanings of the word *live* are: *rest, remain, abide, dwell, to continue, to be permanent, not to perish, to exist.*

So quite literally, to *believe* in Jesus Christ as Creator, Savior and Lord is to have physical existence in Jesus.

> **John 15:4 Abide in me, and I in you. As the branch cannot bear fruit of itself, except it abide in the vine; no more can ye, except ye abide in me.**

We tend to only think of *believing* as something we conjure up in our minds, soul and spirits. However, it is also very much connected to our physical beings—our created beings.

*Believing* is who we are.

When we look at these words written in Scripture, we can begin to understand the importance of the physical act of creation and Jesus as that Creator.

We were created by Jesus, to be saved by Jesus, to live eternally with Jesus. After all, God said in Genesis 1:26:

**Let us make man in our image: after our likeness.**

Plants have a body while animals have both a body and consciousness.

*Let us make man in our image* necessitates human likeness to the triune God. Man was not only made with both a body and a consciousness of the created soul, but also with a third entity, the image of God,

Made in the image of God, we are called to think, and through our understanding of how the world must have been created by God (Heb 11:3), have a reasonable faith which allows us to truly commune and fellowship with the Creator.

*After our likeness* implies that man was created in God's spiritual image, and this could be construed to mean in His physical image as well. Our bodies are indeed created for divine fellowship: erect posture; upward looking and expressive faces; brains, mouths and voice boxes designed for speech.

I know what you're thinking.

*Come on Russ, I believe that Jesus is the Creator. But couldn't He have created over **millions of years**, or couldn't he have used evolution to create? What does believing in **millions of years** have to do with it?*

Accepting God at His Word has EVERYTHING to do with believing, based on logical proof (some seen, some unseen), in the ONE TRUE Jesus.

> **Hebrews 11:3**
> **Through faith we understand that the worlds were framed by the word of God, so that things which are seen were not made of things which do appear.**

Both *millions of years* and *Darwinism* are opposed to what God has told us He has done. For instance, Jesus did not believe that He made mankind *millions of years* after the beginning.

> **Mark 10:6 But from the beginning of the creation God made them male and female.**

Also, the man-made belief of *millions of years* opened the door for the man-made philosophy of *Darwinism*. Together, these non-Scriptural philosophies have misled untold billions of people.

To believe in a Jesus that used *millions of years* is to put your faith in a form of Jesus who is NOT FOUND in the Bible. Don't trust me on this, or anyone else. Look for yourself. You will NOT find an old-earth version of Jesus Christ in God's inspired Word.

And this is what *millions of years* and *Darwinism* are really about. You see, Satan doesn't want you to believe in the Jesus of the Bible, and with good reason:

> **The Jesus of Scripture is the Jesus Who created us, saves us, sanctifies us and glorifies us. He is the seed of the woman who will bruise the head of the serpent.**

Satan desires that people reject their Savior. By punching holes in people's belief in Jesus the Creator—and this is what *millions of years* and Darwinism do—the COS† is corrupted and holes are also punched in Jesus the Savior.

It's just that simple.

Most of our youth have figured this out and many of the young adults who have left the Church still believe in Jesus. However, they see the Church failing to stand up for the one true Jesus and for the Truth of His Word.

Then they see Church leaders, Christian schools and seminaries distorting God's Word in order to force-fit secular beliefs in place of Biblical teachings.

They also see the Church refusing to teach them why they can trust in the God of the Bible.

The result is that they see the Church as being hypocritical and as not relating to the real world. Since they don't need to attend a church in order to find good snacks, meet friends or hear rock music, they leave. The Church is irrelevant to them.

And of the 15% that remain, how many are being led to worship a Jesus not found in God's Word?

Hummm...the *Submerging Church.*

**The Jesus Found in God's Word**

**John 1:3 All things were made by him and without him was not any thing made that was made.**

This is an ardent statement asserting Jesus Christ made everything before His earthly life and redeeming sacrifice on the cross. He was our Creator before He became our Savior.

**Genesis 1:1 In the beginning God created the heaven and the earth.**

God sent us His only begotten Son, Lord Jesus the Christ, the Creator, Who was before the foundations of the world, to earth in human form. Jesus lived as a man that could eat, drink, walk, talk, and leave a historical trail behind that proves he had been on earth.

**Romans 11:36 For of him and through him and to him are all things.**

Jesus created all things, upholds all things and reconciles all things.

This Biblical Jesus is a very real Jesus; One Who can be tested, studied and observed through His Spirit, through His Word, through His life on earth and via His Creation.

**Romans 1:20 For the invisible things of him from the creation of the world are clearly seen, being understood by the things that are made, even his eternal power and Godhead; so that they are without excuse:**

I rented an auditorium at a university to give several of my Creation seminars. After one of the teachings, I held a question and answer session. A man stood up and said, "I'm a Christian and I believe in an old earth. I believe what I believe on faith. In fact, I think the evidence that you present takes away from faith."

I pointed out that God has given us evidence for a reason and that God wants us to be able to clearly see Him and His handiwork through His Creation.

He threw up his hands in disgust and stormed out of the building. The next evening someone drained the oil from my car as I spoke.

————

This episode prompted me to never park in the same place two nights in a row when speaking to a hostile audience and to do some studying on the word *faith*.

The word *faith* is from the Greek word *pistis* (close to the word *pisteuo*—believe). Faith *(pistis)* means assurance or fidelity.

> **Hebrews 11:1 Now *faith is the substance of things* hope for, the *evidence of things* not seen.**
> (emphasis mine)

Now let's look at the word *hope*.

The archaic meaning of *hope* is *trust*. *Trust* is, again, tied to the word *true*.

*Hope* is also related to the word *heap*. *Heap* is to pile up a great quantity. In essence, we pile up a great quantity of *truth*— that is, *HOPE*.

But not just any *truth*.

God's *Truth*.

As Christians, Jesus Christ is our *HOPE*.

AGAIN, *HOPE* is not just wishful thinking, or an optimistic emotion. It is based on *Truth* and in *Truth*.

*Faith* is based on *hope*; however, God also wants us to have solid evidence—both physical and unseen evidence that can be known through reasoning—and He has provided us with lots of it that is faith-supporting.

> **1 Thessalonians 5:21 Prove all things; hold fast that which is good.**

God has provided us with evidence so that our faith will be solid.

**Ephesians 6:16 Above all, taking the shield of faith, wherewith ye shall be able to quench all the fiery darts of the wicked.**

The shield in Roman times was made of metal and it provided a solid wall to protect its bearer. Our *shield of faith* is to be more solid than any man-made device. It is to be based upon hope in the Jesus Christ found in Scripture and supported by the I-beams of His Spirit, His Word and the proof that God has provided to us throughout His Creation.

Compromise only exists due to a lack of faith. A shield made of compromised faith will not deflect cheap insults, much less fiery darts. We certainly have no excuse if our faith is not solid.

In Philippians we are told to have THE FAITH OF CHRIST:

**Philippians 3:9 And be found in him, not having mine own righteousness, which is of the law, but that which is through the faith of Christ, the righteousness which is of God by faith:**

Well, Christ's faith is based on a lot more than just hope. He is the Alpha and the Omega. Jesus' faith is rock-solid as He knows for a fact Who He is. Jesus has a strong, solid, evidence-backed faith which He knows is totally reasonable and totally trustworthy.

We are to *have the faith of Jesus.*

So faith is not just a spiritual gift or something that, if we clench in our fist or grind our teeth over, we can conjure up to get us through our troubles.

No, faith is hope combined with solid, reliable facts and proof.

We need to realize that Jesus is our Creator, Judge, redeeming Savior and Lord.

Thus, when we say we have faith in Jesus Christ we should be proclaiming that we hold to a hope-filled yet solid, evidence-backed fact that Jesus is Who He claims to be, and our faith in Him should be impossible for us to deny.

When the Body of Christ turns its back on the solid evidence and re-invents a Jesus that does not appear in Scripture, a Jesus based on man's opinion about *millions of years* of time, we are denying that Jesus Christ was capable of creating the way He said He did, and in the time He said it took Him to do so.

The bottom line is that we are calling Him a liar and we are denying Who He says He is.

And Jesus' work as Creator is not yet done. He has promised to give us a new heaven and earth and new, glorified, eternal bodies.

> **Philippians 3:21 Who shall change our vile body, that it may be fashioned like unto his glorious body, according to the working whereby he is able even to subdue all things unto himself.**

As omnipotent Creator, Jesus is able to resurrect dead bodies and glorify everlasting bodies which are not corrupted by disease, pain or death.

I am looking forward to that reality.

But think about it for a moment:

If we don't believe that Jesus Christ could create our mortal body at His spoken Word it is hypocritical of us to suggest He can create our immortal body in the blink of an eye.

And people see the hypocrisy within the Church Body.

Jesus, as Creator, creating the way He said He did, in the time He said it took, is INDISPENSABLE to the Church.

Using Scripture as our guide, the earth is a few thousand years old, which is a long period of time.

If the age of the earth is a stumbling block to you or to someone you care about read our report *371 Days That Scarred Our Planet.* (Old-earth beliefs are THE biggest compromise within the Church today.)

This book, as I alluded to earlier in this report, provides the most easy-to-read, follow and understand scientific and Scriptural reasons not to believe in *millions of years* that I have ever come across. Clearly communicating these issues is our goal.

I believe the Church needs to take hold of the reality of a recent, young-earth Creation, made by Jesus Christ our Creator.

Christian leaders need to grasp the reality of the worldwide flood, decreed by our Judge, Jesus Christ.

Christian schools and seminaries need to embrace the scientific evidence that supports Biblical Truth.

The Church Body needs to cease our hypocrisy and start believing God's Word.

The *Submerging Church* needs to *submerge* itself into the uncompromised Word of God and teach people how to defend the true Christian faith.

The Church must rebuild itself upon the rock-solid foundation of Scripture, which is ALL essential, word for word and cover to cover.

Only then can we begin tying Biblical teachings into the world we live in and make the Bible relevant in our own eyes and our own lives. Only then can we make it relate to other people's lives.

Are our origins—as revealed in the Book of Genesis—an essential or a *non-essential* issue? Keep this thought in mind as you read this anecdote and the next chapter.

---

Shortly after I presented *Science Versus Darwinism* in a college auditorium packed with students and faculty, one professor began a blog site attacking me over the worldwide web. He then put together an accredited course titled *Science and Creationism* attacking Biblical Creation and Bible-believing Creationists, me in particular.

In an attempt to evangelize his secular beliefs he e-mailed:

*"I was raised Church of Christ and then attended a Calvary Chapel. My father was a Bible college professor. I even led Bible studies and converted scores of young people to evangelical Christianity."*

However, he claimed that he lost his faith after studying old-earth geology in college and had since misled hundreds of Christian kids. He added that he planned to mislead hundreds more.

The matter of our origins was essential to this former Believer and to each of the Christian-raised students he has misled already.

The matter of our origins is also essential to those he will cause to stumble in the future – even though they are likely unaware of it.

# FIVE

# Word for word,
# cover to cover

If you've read our books or watched our DVDs you know that our ministry is based on the fact that God's Word, the Holy Bible, is true…

**word for word, cover to cover.**

As I write this chapter I'm fresh off a three-week speaking tour that took my wife Joanna and me to churches in Iowa, Colorado and New Mexico.

It's a great pleasure to be used by God to reveal the relevance and Truth of His Word to people who are seeking Him. Furthermore, as I see the scales fall off the eyes of folks who have had their faith weakened by false secular teachings, and I observe them becoming unashamed to believe in, teach and preach the whole Bible – and not simply dismiss the first eleven chapters of Genesis as being *non-essential* – I get a glimpse of the power of God's Word firsthand.

Shortly before that trip we worked with the pastors of six churches in an area of small communities in Arizona. Afterward I got a call from one of those pastors. He said our visit was the main subject at their next pastors' meeting:

**Especially how we impacted the pastors.**

Some who had thought, or bordered on thinking, that our Genesis-based teaching might be a *non-issue* now have their eyes opened. Now these pastors and their congregations know with confidence they can trust not only the Book of Genesis, but the entire Bible as written. That is, they know they can build their house upon the rock of God's uncompromised Word.

### Big impact!

This showed me that **many of the church doors that are closed to our teachings are shut due to ignorance** of the importance, the relevance, of the issues we cover.

Many pastors, church leaders, church members and seekers fall into a category I call *the unknowingly deceived.* All they need is a bit of Biblically-based nudging provided through a clear presentation of the foundational issues laid down in Genesis to become convinced of the relevance of Creation, evolution and age-of-the-earth issues.

I strongly identify with these people, as I was unknowingly deceived from as early as the third grade right on through my college years and beyond.

Over and over I heard how it took *billions of years* for the universe to form all on its own; how dinosaurs had been extinct for *millions of years*; or how Grand Canyon had taken *millions of years* to form – and the Church provided me with nothing to combat these secular interpretations. So I mixed these worldly teachings into my belief of who Jesus was and what He did or did not do.

As I previously explained, even after I made a profession of faith and asked Jesus into my heart as my personal Lord and Savior I still clung to the secular brain-washing and became a Theistic Evolutionist.

I certainly thought that I was a Bible-believing Christian. But ask yourself this question: Did I really believe all the Bible had to say about Who Jesus was and what He is capable of doing?

I can only thank God that He allowed me to see the Truth through the teachings of dedicated Creation apologists.

As the scales fell from my eyes, I began to see the TRUTH of ALL of God's Word.

Because of my personal experience, *my time in the desert*, I understand why so many professing Christians question the authority of God's Word, almost always starting with the first five words of Scripture:

### Genesis 1:1a In the beginning God created...

The primary purpose of Creation, Evolution & Science Ministries is to build the Biblical foundation of Creation for well-intended folks within the Church who have stumbled and trust in a Jesus who used evolution or *millions of years* of death to get us here.

We rebuild the foundation so people can trust in the one true Christ as described to us in His Word.

Because of my passion for presenting the Biblical foundations laid down in the Book of Genesis, I'm often asked if belief or non-belief in Biblical Creation is a salvational issue. This includes the age-of-the-earth and global flood issues as well—did the flood really happen as is reported in Genesis and as referred to many times throughout the Bible, and does it have a bearing on salvation?

The answer to each question is *I do not know*. And neither does anyone else. Such a judgment is left to our Lord and Savior Jesus Christ alone to make. .

Whether or not it is required for salvation, a person's acceptance or rejection of these issues does have a direct impact on what they believe. For instance, if Jesus is not telling us the truth about His past judgment (the flood), why believe Him about a future one? And if we cannot believe in a future judgment, from what will we need saving?

Necessary for salvation, maybe, maybe not; necessary for the doctrines of salvation, yes.

### Matthew 7:1-5
**1 Judge not, that ye be not judged.**
**2 For with what judgment ye judge, ye shall be**

**judged: and with what measure ye mete, it shall
be measured to you again.**
**3 And why beholdest thou the mote that is in thy
brother's eye, but considerest not the beam that
is in thine own eye?**
**4 Or how wilt thou say to thy brother, Let me
pull out the mote out of thine eye; and, behold, a
beam is in thine own eye?**
**5 Thou hypocrite, first cast out the beam out of
thine own eye; and then shalt thou see clearly to
cast out the mote out of thy brother's eye.**

Trust me when I tell you that, as a former Theistic Evolutionist, I had to remove a large beam from my eye before God would allow me to assist others in draining the mote filled with secular teachings that is clouding their sight.

While believing in Biblical Creation may or may not turn out to be salvational in and of itself, believing in the one true Jesus Christ IS salvational. He is the ONLY way to the Father.

**John 14:6 Jesus saith unto him, I am the way,
the truth, and the life: no man cometh unto the
Father, but by me.**

So what the eternal question of salvation boils down to is this:

**Does it really matter if we believe in Who Jesus
is and in what He claims He has done as long as
we accept salvation through Him?**

Please consider the following with an open mind.

I have heard many renowned Creationists state that what a person believes about Creation is not an issue of salvation and I feel this could be a grave error on their part. The Bible is very clear about how salvation is achieved. We discussed that in the previous chapter.

But there is another aspect of the salvational question that I must confess causes me to wonder where God might draw the line on who truly is saved and who isn't saved. We will get to this more in depth in the next chapter when we talk about Believers, Non-Believers and Make-Believers.

With regard to the questions above, and whether these are salvational issues, I believe that the Biblically-correct response is an honest:

> *I do not know;* **however, I believe that Biblical compromisers are rolling the dice with their eternal destiny.**

How so?

Let's look to the Word of God for our answer.

In Exodus we are given the Ten Commandments, written by the very finger of God Himself.

The first Commandment is to have no other gods besides the one true God – Who is found in Scripture.

> **Exodus 20:3 Thou shalt have no other gods before me.**

The second Commandment is that we are not to make up any graven images (false gods) and bow down to them.

> **Exodus 20:4-5**
> **4 Thou shalt not make unto thee any graven image, or any likeness *of any thing* that is in heaven above, or that is in the earth beneath, or that is in the water under the earth:**
> **5 Thou shalt not bow down thyself to them, nor serve them: for I the Lord thy God *am* a jealous God, visiting the iniquity of the fathers upon the children unto the third and fourth *generation* of them that hate me;**

Psalm 96:5 and Jeremiah 10:11-12 show that the reason our God is different than pagan gods or idols is because He is the Creator.

An idol is a man-made image of a false god. Idolatry is the worship of such an idol. Thus making up a form of god who is not described to us in God's own Word is, I believe, rolling the dice.

Then in the middle of His Ten Commandments, while discussing the Sabbath Day, we are reminded:

> **Exodus 20:11 For in six days the Lord made heaven and earth, the sea, and all that in them is, and rested the seventh day: wherefore the Lord blessed the Sabbath day, and hallowed it.**

Keep in mind that this is God's own description of what He has done.

Again, the first five words of the Bible are *In the beginning God created.* God claims to have created in the beginning, not over long ages of time.

> **Matthew 19:4 And he answered and said unto them, Have ye not read, that he which made them at the beginning made them male and female,**

Jesus, our Creator, claims to have made mankind in the beginning as opposed to *billions of years* after the beginning.

> **Revelation 4:11 Thou art worthy, O Lord, to receive glory and honour and power: for thou hast created all things, and for thy pleasure they are and were created.**

Jesus—not Satan—deserves the glory and the honor for His creative works.

**Romans 5:12 Wherefore, as by one man sin entered into the world, and death by sin; and so death passed upon all men, for that all have sinned:**

The eternal God is infinitely good. It is only logical that the world He created would reflect His very nature. He could not plausibly create a world with death. Death entered God's perfect creation following man's original sin. *Billions of years* of death is NOT what the one true God used to bring man into existence.

**1 Corinthians 15:26 The last enemy *that* shall be destroyed *is* death.**

Death is an enemy to God's Creation, not His ally in creating our world and universe.

**Genesis 7:19 & 21**
**19 And the waters prevailed exceedingly upon the earth; and all the high hills, that were under the whole heaven, were covered...**
**21 And all flesh died that moved upon the earth, both of fowl, and of cattle, and of beast, and of every creeping thing that creepeth upon the earth, and every man:**

The inspired Word of God claims that our righteous Savior judged man's sin with a global flood.

The proof of this is that the crust of the earth is primarily made of sedimentary layers of rock comprised of sediments which were laid down by water.

During the global flood all the flesh died that moved upon the earth, including fowl, cattle, beast and every creeping thing. The flood victims were then buried in the stratified layers to become today's fossil record.

Old-earth beliefs are based on the earth's sedimentary layers of rock forming at a gradual and uniform rate over long ages of time. That scoffers will make this claim is foretold in 2 Peter 3:3-4.

> **3 Knowing this first, that there shall come in the last days scoffers, walking after their own lusts, 4 And saying, Where is the promise of his coming? for since the fathers fell asleep, all things continue as they were from the beginning of the creation.**

Yes, a global flood destroys old-earth beliefs by explaining how the water-borne layers formed quickly and recently as a direct result of God's watery judgment.

This leaves old-earth followers with two choices:

1] humble themselves and accept God's Word; or
2] deny the worldwide flood.

Option number two is prophesied in 2 Peter 3:5-6 which tells us that scoffers will choose to deny the global flood.

> **5 For this they willingly are ignorant of, that by the word of God the heavens were of old, and the earth standing out of the water and in the water: 6 Whereby the world that then was, being overflowed with water, perished:**

Scoffers will deny the global flood because such a worldwide judgment would wash away their old-earth beliefs.

Scripture also prophesies that people will claim to believe in a form of god but deny the power of that deity during the last days.

> **2 Timothy 3:1-5 1 This know also, that in the last days perilous times shall come.**

**2  For men shall be lovers of their own selves, covetous, boasters, proud, blasphemers, disobedient to parents, unthankful, unholy,**
**3 Without natural affection, trucebreakers, false accusers, incontinent, fierce, despisers of those that are good,**
**4 Traitors, heady, highminded, lovers of pleasures more than lovers of God;**
**5  Having a form of godliness, but denying the power thereof: from such turn away.**

God's inspired Word warns us to turn away from those who claim to believe in God but deny He has the power He claims to have.  Hmmmm….from such **TURN AWAY.**

Jesus Christ Himself refers to the global flood as an historical fact.

**Matthew 24:37-39**
**37 But as the days of Noe *were*, so shall also the coming of the Son of man be.**
**38 For as in the days that were before the flood they were eating and drinking, marrying and giving in marriage, until the day that Noe entered into the ark,**
**39 And knew not until the flood came, and took them all away; so shall also the coming of the Son of man be.**

So again, is making up a Jesus who is NOT found in Scripture, a Jesus who used evolution or *millions of years* of death to create us, a salvational issue?

I leave the judgment up to Jesus Christ.  I feel called to present God's Word and the observable facts supporting Scripture so you can make an informed decision for yourself.

But I NEVER tell folks it is not a salvational issue. NEVER.

So what about well-intentioned seekers who have honestly been misled by old-earth beliefs that are taught within the *Submerging Church*? After all, 90% of accredited Christian schools and most seminaries are teaching various old-earth philosophies; doesn't that make believing them all right?

> **2 Thessalonians 2:3 Let no man deceive you by any means: for that day shall not come, except there come a falling away first, and that man of sin be revealed, the son of perdition;**

We are warned not to let others deceive us and that there will be a falling away from the Truth in the end times.

Jesus also warned us to beware of those who appear to be on His side but take us away from His Truth.

> **Matthew 7:15 Beware of false prophets, which come to you in sheep's clothing, but inwardly they are ravening wolves.**

Keep in mind that God always leads us to His Word, not away from it.

*Well, for crying out loud!* you might be thinking.

*There are so many wolves disguised as sheep today, pointing us to what this guy wrote or to what that other person said; how in the world can we tell a friendly sheep from a ravenous carnivore?*

The answer is that we are supposed to follow God's Word. The Holy Spirit will shepherd us along the narrow path that leads to heaven as long as we put our faith in God's Word.

The Holy Spirit will always lead us to God's Word, never away from it.

Jesus tells us that we are to tell good from evil, the sheep from the wolves, by the results they produce.

**Matthew 7:16-20**
**16 Ye shall know them by their fruits. Do men**
**gather grapes of thorns, or figs of thistles?**
*17* **Even so every good tree bringeth forth good**
**fruit; but a corrupt tree bringeth forth evil fruit.**
**18 A good tree cannot bring forth evil fruit,**
**neither can a corrupt tree bring forth good fruit.**
**19 Every tree that bringeth not forth good fruit**
**is hewn down, and cast into the fire.**
**20 Wherefore by their fruits ye shall know them.**

With billions of people having rejected Jesus Christ as their Lord and Savior due to old-earth beliefs, shouldn't we have seen the rotten fruit and rejected *millions of years* as a corrupt tree?

Old-earth beliefs opened the door for Darwinism which has caused untold millions more to stumble and lose their testimony.

*Millions of years leading to Darwinism* has opened the door to the drug culture, the sexual revolution, the skyrocketing number of sexually-transmitted diseases and the abortion of 50 million babies in America alone...and this is only a short list of the evil fruit sprouting from these corrupt trees.

According to directions given to us by Jesus Christ Himself, we are supposed to tell good from evil by the fruit. Yet with more than 90% of accredited Christian schools and the majority of seminaries teaching various old-earth beliefs, we are not doing very well at discerning good from bad, are we?

Granted, many have been honestly deceived just as I once was. But many of those who've been deceived might not have stumbled had they read their Bibles and placed their faith in God's Word.

Ultimately we are responsible for what we decide to believe or not believe. And while only Jesus can judge where we each stand as servants to Him, He warns us that many who consider themselves to be saved will be in for a terrible shock on THAT DAY.

**Matthew 7:21-23**
**21 Not every one that saith unto me, Lord, Lord,**
**shall enter into the kingdom of heaven; but he**
**that doeth the will of my Father which is in**
**heaven.**
**22 Many will say to me in that day, Lord, Lord,**
**have we not prophesied in thy name? and in thy**
**name have cast out devils? and in thy name done**
**many wonderful works?**
**23 And then will I profess unto them, I never**
**knew you: depart from me, ye that work**
**iniquity.**

Those people rolled the dice.

And lost.

Those people did not have the salvation they thought they had.

In the days of the great prophet Elijah, the ancient Israelites were compromising their worship of the one true God with pagan beliefs and practices. This was not acceptable to God so Elijah exhorted the people:

**1 Kings 18:21 And Elijah came unto all the**
**people, and said, How long halt ye between two**
**opinions? If the Lord *be* God, follow him:**

So how do we know today Who the real Jesus is so we can follow Him and not stumble after some man-made, false version of Jesus?

**We pray for Jesus to guide us by His Holy Spirit**
**to trust in Him as we read about Him in His**
**Word. We critically examine scientific evidence**
**to see if it fits into a Biblical interpretation.**
**Then, with understanding and the confidence**
**that comes with knowledge, we *believe* in the**

**Jesus found in the Bible with a faith that is never blind, but fully reasonable.**

It really is quite simple. Not always easy, but simple. Choosing to believe in the Biblical Jesus may cause us to lose friends and family, jobs or other opportunities, and eventually perhaps even our earthly lives, but all the more is our eternal gain.

The Word of God claims that the one true Jesus Christ is our Creator, Judge and redeeming Savior. And all He asks is that we believe in Him, calling on Him to save us.

**John 3:16 For God so loved the world, that he gave his only begotten Son, that whosoever believeth in him should not perish, but have everlasting life.**

Thus, when I am asked if belief or non-belief in Biblical Creation is a salvational issue my sincere answer is that *I do not know*. But my heartfelt advice is this: Read about and believe in the Jesus Christ found in God's uncompromised Word, word for word and cover to cover.

This is the one sure way that you will spend eternity in heaven with the one and only true begotten Son of God.

And always keep in mind this admonition from the apostle Paul:

**Romans 12:2 And be not conformed to this world: but be ye transformed by the renewing of your mind, that ye may prove what is that good, and acceptable, and perfect, will of God.**

After speaking at a college campus several associate pastors from a nearby church came up to me to tell me how blessed the folks in their church would be to hear my teachings.

The leaders of the church's twenty-to-thirty-year-old group told me that Darwinism was the main reason cited by those leaving his group and he could not wait to get me to visit.

They soon discovered that the senior leadership of the church where they were employed believed in old-earth versions of Christ. I would not be allowed to help their flock.

They found themselves at a crossroads of life and a choice had to be made: shepherd or hireling?

> John 10:11-13
> 11 I am the good shepherd: the good shepherd
> giveth his life for the sheep.
> 12 But he that is an hireling, and not the
> shepherd, whose own the sheep are not, seeth the
> wolf coming, and leaveth the sheep, and fleeth:
> and the wolf catcheth them, and scattereth the
> sheep.
> 13 The hireling fleeth, because he is an hireling,
> and careth not for the sheep.

Each chose to flee from the issue and remain employed by this large church, sitting at the heads of the banquet tables while leaving their sheep to fend for themselves against the ravenous wolves of the world.

# SIX

# Believers, Non-Believers and Make-Believers

In our ministry we closely identify with the statement on the back cover attributed to Martin Luther. The *Submerging Church* is fast sinking because it has *flinched* at *In the beginning God created,* the very point which the world and the devil himself are presently, and have been for almost 200 years, attacking.

Not only has much of the Christian Church swept Creation, the Global Flood and the other foundational teachings of Genesis under the rug but much of the *Submerging Church* has been actively aiding and abetting the enemy.

**"The hardest thing to convince a liberal apostate church of today is that the Word of God is true."**

*John MacArthur*

I repeatedly experience the difficulty John MacArthur cited in getting today's *Submerging Church* to believe that God's Word is true, word for word and cover to cover.

Often our enemy disguises himself in order to gain our trust. Once inside of our defenses he causes us to stumble from the narrow path and onto the broad way that leads to our eternal destruction.

**The Bible is rife with warnings about false teachers.**

**2 Corinthians 11: 13-15**
**13 For such are false apostles, deceitful workers, transforming themselves into the apostles of Christ.**

**14 And no marvel; for Satan himself is
transformed into an angel of light.
15 Therefore it is no great thing if his ministers
also be transformed as the ministers of
righteousness; whose end shall be according to
their works.**

The attack from within the Church Body should not come as a surprise to us as Jesus warned us many times to be aware of the *tares among the wheat* and *the goats among the sheep*, and through other parables.

> **Matthew 13:24-28a**
> **24 Another parable put he forth unto them,
> saying, The kingdom of heaven is likened unto a
> man which sowed good seed in his field:
> 25 But while men slept, his enemy came and
> sowed tares among the wheat, and went his way.
> 26 But when the blade was sprung up, and
> brought forth fruit, then appeared the tares also.
> 27 So the servants of the householder came and
> said unto him, Sir, didst not thou sow good seed
> in thy field? from whence then hath it tares?
> 28a He said unto them, An enemy hath done this.**

*While men slept*...**while men were not paying attention.**

Martin Luther called this Bible passage *the parable of the tares which an enemy sowed in the field.* As the terms *tares and wheat* appear often in this report it is important you know what the teller of the parable, Jesus the Christ, was telling his audience.

Questioned by His disciples as to the meaning of the parable, Jesus said:

**Matthew 13:37-39**
**37 He answered and said unto them, He that**
**soweth the good seed is the Son of man;**
**38 The field is the world; the good seed are the**
**children of the kingdom; but the tares are the**
**children of the wicked** *one***;**
**39 The enemy that sowed them is the devil; the**
**harvest is the end of the world; and the reapers**
**are the angels.**

The wheat are the Believers and are the result of the good seed, while the tares are Satan's minions who corrupt what the field produces. The tares are sown by the Great Deceiver.

The wheat and the tares are in direct competition for the hearts and minds of true seekers. The weeds soak up the resources of the Church Body. Non-believers push folks away from their loving God. Non-believers plant the seeds of doubt which produce the fruit of compromise. Make Believers go along with the flow, not caring what anyone truly believes. And the harvest will be the end of the world.

**The rest of the parable**

**Matthew 13:28b-30**
**28b The servants said unto him, Wilt thou then**
**that we go and gather them up?**
**29 But he said, Nay; lest while ye gather up the**
**tares, ye root up also the wheat with them.**
**30 Let both grow together until the harvest: and**
**in the time of harvest I will say to the reapers,**
**Gather ye together first the tares, and bind them**
**in bundles to burn them: but gather the wheat**
**into my barn.**

Which are the wheat and which are the tares?

This may not be readily apparent until the final harvest so we are to leave that judgment up to Jesus Christ who will send His angels to righteously divide the two.

Martin Luther expressed the opinion that Jesus spared the tares to give them every possibly opportunity to confess their sins, repent and transform into wheat. Certainly God does not want anyone to perish but all to come to repentance.

Still, God's patience will one day come to an end. His angels will bundle up the tares to be burned while sending the wheat—the true Believers—on to spend eternity with Jesus Christ in heaven.

If we are to be honest with one another we must acknowledge that every church is made up of Believers, Non-Believers and Make-Believers—the wheat and the tares. Those who are in the categories of Non-Believers and Make-Believers fall into two major divisions:

- The ignorant and
- the willingly deceived.

The former can be *cured* through our teachings. The latter can only be cured if God elects to soften their hearts via the Holy Spirit.

There are many people who try to follow Jesus but their hearts are never truly with Him. These folks tend to go to church as a social function as opposed to seeing it as a time of worship and a chance to grow in Biblical knowledge. Although they go through the motions, they are not truly seeking Jesus Christ to bring Him glory and make Him the Lord over their lives.

**Matthew 25:2-12**
**2 And five of them were wise, and five *were* foolish.**
**3 They that *were* foolish took their lamps, and took no oil with them:**
**4 But the wise took oil in their vessels with their lamps.**

**5 While the bridegroom tarried, they all slumbered and slept.**
**6 And at midnight there was a cry made, Behold, the bridegroom cometh; go ye out to meet him.**
**7 Then all those virgins arose, and trimmed their lamps.**
**8 And the foolish said unto the wise, Give us of your oil; for our lamps are gone out.**
**9 But the wise answered, saying, *Not so*; lest there be not enough for us and you: but go ye rather to them that sell, and buy for yourselves.**
**10 And while they went to buy, the bridegroom came; and they that were ready went in with him to the marriage: and the door was shut.**
**11 Afterward came also the other virgins, saying, Lord, Lord, open to us.**
**12 But he answered and said, Verily I say unto you, I know you not.**

The ten maids were supposedly all awaiting Jesus. However, five of them were only going through the motions. Those five were not truly expecting Him to return and when He came Jesus told them that He did not know them.

It's interesting that studies say half of today's Christians believe in a Christ who used *millions of years* of death to bring man into the world.

Such a Christ is not found in God's Word.

The denial of the one true Jesus Christ has been an issue even from the early days of the Church.

**Philippians 3:17-18**
**17 Brethren, be followers together of me, and mark them which walk so as ye have us for an ensample.**

**18 For many walk, of whom I have told you
often, and now tell you even weeping, *that they
are* the enemies of the cross of Christ:**

False teachers were a problem for the Church within twenty-five years of its founding. We are without an excuse for allowing them to misguide us today, 2,000 years later.

**Philippians 3:19 Whose end *is* destruction, whose
God *is their* belly, and *whose* glory *is* in their
shame, who mind earthly things.**

They *mind earthly things* such as man's opinions rather than God's clear Word.

Old-earth beliefs and Darwinism, which are in direct conflict with God's Word, are such earthly things.

Again, the people fooled into accepting man's ever-changing ideas over God's never-changing Word fall into two major divisions: the ignorant and the willingly deceived.

**Titus 1:16 They profess that they know God; but
in works they deny *him*, being abominable, and
disobedient, and unto every good work
reprobate.**

Non-Believers and Make-Believers may claim to know the one true Jesus but they will actually deny Him when it comes right down to it. They deny God's power and authority by accepting man's anti-Scriptural opinions over the very Word of the God they claim to worship. This applies directly to those who blend secular beliefs of *millions of years* or Darwinism into the Bible.

Such compromised teachings are reprobate as they lead people away from trusting God's Word.

Many Make-Believers grow from the seed planted among the thorn bushes. They started out on the right path but let the temptations of the world lead their heart away from the Lord. Though they produce no fruit they still attend church.

Our teachings often help Make-Believer Christians free themselves from the thorn bushes of secular teachings which are choking out their faith. Clearing away such tares allows them to strive ahead in their quest to stay on that narrow way.

For other Make-Believers, our teachings are like the seed sown on rocky soil. They refuse to submit their lives to the solid foundational root system and continue living in the broad ways of the secular world.

I believe that *make-believing* in our faith can be so subtle, especially with all of the false teachings accepted inside of the *Submerging Church*, that many well-meaning seekers have actually become Make-Believers—and they do not even recognize it.

They think they are doing just fine but they are honestly ignorant of the fact that they have been deceived. Our teachings can bless these folks immensely – if we can get our teachings to them.

Here is a message I received from one such victim of the *Submerging Church*.

### "I am beginning to doubt Jesus exists but I am saved because I was baptized."

Salvation comes by turning from sins (repentance) and trusting in Jesus (faith). Both of these appear to be lacking in this victim of the *Submerging Church*. I fear there are many more people who operate along this same line of thought than we realize.

The lack of faith openly accepted, and even promoted, within the *Submerging Church* leads many people to assume one can believe or not believe anything they like and still be saved.

**Matthew 24:11-12**
**11 And many false prophets shall rise, and shall deceive many.**

**12 And because iniquity shall abound, the love of many shall wax cold.**

The seeds of doubt are flourishing within the *Submerging Church* and the love of many has indeed turned cold.

Satan is subtle which is why he is so successful, and one of his best efforts is aimed at deceiving people into worshipping a false Christ. One way he does this is to take a few verses of Scripture out of context to seemingly support a handful of lies—and through his subtlety he deceives many.

**2 Thessalonians 2:3**
**Let no man deceive you by any means: for *that day shall not come*, except there come a falling away first, and that man of sin be revealed, the son of perdition;**

Here we are warned the Church will eventually fall away, that it will *submerge.*

Remember, the one sure way to keep from being deceived is to spend time reading the Bible and to compare all teachings to God's Word to guide us along that narrow path.

**Revelation 2:4 Nevertheless I have *somewhat* against thee, because thou hast left thy first love.**

As Christians we need to set our hearts and minds to growing in the one true Jesus. We are to strive to grow and outpace the thorn bushes which desire to choke out our first love which is our faith in Christ.

Non-Believers, and their allies within the *Submerging Church,* attempt to make the Church Body conform to this world. (Consider what is said in Romans 12:2.) And the god of this world is Satan.

The falling away is well underway.

The growing tilt to the left of the *evangelical Christian Church* is rapidly leading more and more Christians into becoming indistinguishable from the secular left.

Many of the 15% of Christian-raised kids who do not leave the church altogether by the age of twenty often fall prey to these apostates who seem to think that conformity will win converts.

History reveals that conformity churns out weak Believers and Make-Believers who are poor Christian witnesses. Conformity also leads to hypocrisy within the Church which leads many people to leave the *Submerging Church* altogether.

We are sternly warned to beware of such conformity and to examine our own hearts as well.

> **Hebrews 3:10-12**
> **10 Wherefore I was grieved with that generation, and said, They do alway err in *their* heart; and they have not known my ways.**
> **11 So I sware in my wrath, They shall not enter into my rest.**
> **12 Take heed, brethren, lest there be in any of you an evil heart of unbelief, in departing from the living God.**

So how in the world do we recognize apostate teachings in others as well as within our own understanding? We compare all teachings to the Word of God. Anything taking us away from God's Word is not from Him.

> **2 Thessalonians 3:6 Now we command you, brethren, in the name of our Lord Jesus Christ, that ye withdraw yourselves from every brother that walketh disorderly, and not after the tradition which he received of us.**

The Word of God tells us to take ourselves away from those who do not follow the ways of God.

Still, I know many Christians who knowingly attend churches whose leadership compromises Jesus Christ with various man-made versions of Him. They financially and emotionally support these false leaders rather than seeking out and supporting a God-honoring pastor.

> **Ezekiel 33:31-32**
> **31 And they come unto thee as the people cometh, and they sit before thee *as* my people, and they hear thy words, but they will not do them: for with their mouth they shew much love, *but* their heart goeth after their covetousness.**
> **32 And, lo, thou *art* unto them as a very lovely song of one that hath a pleasant voice, and can play well on an instrument: for they hear thy words, but they do them not.**

I know many God-honoring pastors with very small congregations who could be so encouraged if more Christians would support them. But folks seek out teachers who will scratch their itching ears. It is of little wonder the Church is *submerging.*

This is a dangerous place for the Church Body to be and false religious leaders are allowing them to think they are fine. Jesus gave a stern warning:

> **Matthew 12:30 He that is not with me is against me; and he that gathereth not with me scattereth abroad.**

Please, look in a mirror and honestly ask yourself:
Am I a Believer, Non-Believer or Make-Believer?

What about the leadership of the church you attend?

I'm standing outside of a church in Maryland where I am about to give my *Noah's Ark and Dinosaurs* teaching to a youth group. As I wait I am also observing the gathering youth…and I feel sad for them.

These are Christian-raised youth, standing outside of a Christian church and quite frankly they look as if they are auditioning for an MTV video.

Most of them, including the youth leaders, have multiple tattoos. The clothing is tight and suggestive on the girls and baggy and revealing on the boys.

Most of these kids have been educated in public schools and their church has failed to give them a reasonable defense of God's Word. Even worse, many have seen Christians compromise Jesus' earthly teachings with opposing secular beliefs.

They see the hypocrisy when the Church preaches about the Bible's moral values as if those parts of the Bible are true. As a result they do not see the Bible as being relevant to their lives.

They are victims of the *Submerging Church*.

I can't help but to be mad about it.

## SEVEN

# Hypocrisy Abounds

On one hand the *Submerging Church* is desperately trying to pander to secular teachings. Then, on the other hand, it will boldly proclaim the need for people to live a Christ-centered life, and guess what?

Youth, among others, are seeing the hypocrisy.

Preaching that the Bible gives you the *rules for right living* and then denying that the Bible means what it clearly says about Creation-related issues has serious consequences.

For instance, one in five abortions is performed on women who profess to be born-again, evangelical Christians. *(Guttmacher Institute Study)*

In fact, the statistics for Christians committing fornication, adultery, and getting divorced are virtually the same, and in some cases higher, than non-Christians.

### What is happening?

It is primarily an issue of conflicting worldviews as our secular society has taught the past fifty years of American children that they evolved over long ages of time without God.

However, the dagger through the heart of our society is that the Church has failed to provide a defense for the trustworthiness of God's Word.

By compromising the Bible with secular teachings, the Church has basically told the world, including churchgoers, that you can't trust what the Bible has to say.

Compromised, old-earth positions say loud and clear that the Bible's accounts of Creation, how sin and death entered the world, the global flood, Scripture's historical time-line and more are not trustworthy.

And if you can't trust God's Word in these earthly areas, why should we live our lives by what the Bible has to say in other areas?

Such hypocrisy does not fly with people, including today's youth.

Kids see the Church's lack of faith from a very early age (some even before junior high) and studies prove that most Christian children are only waiting until they leave their parents' house before they stop going to church.

The *Submerging Church* seems to be blind or unconcerned as to the extent of the damage non-Biblical teachings on origins have done to the Christian faith and to our society in general.

### What's church got to do with it?

One study found that nearly 89% of alcoholics said they lost interest in religion during their youth. Among young people, the importance of religion is the single best predictor of substance-abuse patterns.

There is also a strong correlation between attending church and the avoidance of crime. Harvard professor Richard Freeman discovered that regular church attendance is the primary factor in preventing African-American urban young people from turning to drugs or crime.

Several studies have found that high levels of commitment to attending church correlate with lower levels of depression and stress.

These polls reveal that people who do not attend church are four times more likely to commit suicide than are regular church attendees. In fact, lack of church attendance correlates more strongly with suicide rates than any other factor, including unemployment.

Church attendance even affects mortality rates.

Studies have shown that going to church can lower blood pressure and enhance survival after a heart attack. For men who attend church frequently, the risk of dying from heart disease is only 60% of that for men who attend infrequently. (David B and Susan Larson, *The Forgotten Factor in Physical and Mental Health*, National Institute for Healthcare Research, Rockville, MD.)

These statistics hold true for other diseases as well. Science seems to be corroborating Scripture (yet, again). But it is not just church attendance that matters, but also what the person believes.

> **Proverbs 10:27 The fear of the Lord prolongeth days; but the years of the wicked shall be shortened.**

To further that point, Dr. Dale Matthews (*The Faith Factor: Proof of the Healing Power of Prayer*, New York; Viking Press) documented studies in which people prayed for patients with arthritis.

To avoid a placebo effect, the patients were told if they were or were not being prayed for. The prayed-for patients' recovery rate was markedly higher than those who were not prayed for.

Interestingly, David Larson, President of the National Institute for Healthcare Research, has found that benefits accrue only to those who practice their faith, not to those who merely profess it. His studies (Larson and Larson, *The Forgotten Factor*) have found it to be extremely unhealthy to hold strong religious beliefs without practicing them.

The most wretched person of all is the one who knows the truth yet chooses to ignore it.

> **Romans 1:21 Because that, when they knew God, they glorified him not as God, neither were thankful, but became vain in their imaginations, and their foolish heart was darkened.**

Marriage and family life benefit greatly from regular church attendance. One study found that church attendance is the most important predictor of marital stability. (Howard M. Bahr, *Journal of Marriage and Family, Religion and Family in Middle America*)

The *Submerging Church* is sending out mixed messages about the believability of the Bible. For the 15% of young people that stay, there is much confusion and uncertainty about Scripture. Their compromised beliefs lead many of them to compromised living. As it says in James 1:8—

**A double minded man is unstable in all his ways.**

Simply stated, the *Submerging Church* does great damage to people seeking the Truth through the Bible.

But not to worry.

Government programs, private non-profits and public schools prove that a person can live a moral and healthy life without the Biblical God.

Right?

**WRONG. Dead wrong.**

For starters, let me clarify one thing. American public schools are a government program. They are one and the same. So I will address them as such.

As of 1962 our public schools were ranked among the very best in the world. We were especially strong in math and the sciences. Then during 1963 Biblical Creation and prayer were kicked out of our public schools, so that our children could be *better educated* in the modern-day beliefs of Darwinism and *millions of years.*

The direct result is that today's American high school seniors are among the worst-educated youth in the world. In one study covering twenty-one nations, American students placed nineteenth in math and science, and dead last in physics.

Even more upsetting is the decline in moral instruction. Teachers have replaced tried-and-true methods of teaching with *self-esteem* classes and *diversity* training. Educators take as many courses on positive reinforcement and social equity as they do on the subject they were hired to teach.

No wonder we score so low on tests. At least we make kids feel worthy while we fail to give them a proper education.

Modern educational techniques, which have literally thrown our children into chaos and destruction, are designed to support the faulty secular worldview.

Children are not taught they were *created* by God, but that they evolved through *millions of years* of mutations. They are not sinful because right and wrong depends on your individual point of view. There was no Original Sin and Redemption does not come through Jesus Christ; it comes through social engineering, proper teaching methods and political correctness.

This has led to the theft of our great Christian and American heritage.

Most youth today know more about gay rights than where Washington D.C. is on a map. In fact, three out of four high schools graduates do not even know who presented the Gettysburg Address.

I cover the tragedy of our rewritten American history in *The Theft of America's Heritage*. Attack and destroy Biblical Creation and you leave a path of annihilation a mile wide that affects not only the Church, but society as well.

The failure of American public education is not because of poor teaching or lack of money. It is due to being based on the secular worldview.

We no longer teach that we are created in the image of God.

We no longer teach that we all have the capacity to sin and be rebellious.

And we no longer teach that redemption comes through Jesus Christ rather than via self-esteem classes.

Unfortunately the *Submerging Church*, once again afraid of rocking the boat and upsetting the secular community, has capitulated on the humanistic indoctrination that is taking place in our schools. In most cases they go as far as to accept or condone secular ideas over the Word of the God they claim to worship.

Pure hypocrisy.

I am constantly asked to go into public colleges to show the science that backs Biblical Creation and refutes Darwinism. I do so when I am asked and supported, but the truth is, the people in the *Submerging Church* need what I teach every bit as much as do the kids in our secular institutions.

Face it, when Christians do not even believe the Bible, why should a Secular Humanist?

## The Bigger Picture

Often I hear someone say, "You can't tell the difference between a Christian and a non-Christian anymore." Sad to say, most of the time the statement is true.

Here's an example:

A friend of ours had worked at a cabinet shop for ten years and paid his family's health insurance directly out of his paycheck.

For two years his wife had been battling cancer, going in and out of treatments. After six months between chemotherapy sessions she returned to start another round of treatment only to learn that her husband's boss, a self-proclaimed Christian, had canceled their health insurance two months earlier and had been pocketing the insurance money taken out of her husband's paycheck.

Our society is in a moral freefall and unfortunately, you often can't tell the Christians from the non-Christians. Postmodernism has hit the *Submerging Church* and relativism has taken away the Bible's guidelines for right and wrong, good and bad. We no longer say *that's wrong*; now we say *that's just wrong for you*.

By flinching on the Biblical worldview, the COS†, which includes a recent, young-earth Creation, the *Submerging Church* has become little different than a social club.

The chasm between what a Christian says in church and what a Christian does in private is directly related to whether that person, and usually the church they attend, stands firm for the authority of God's Word or accepts secular opinions over the Word of God.

Compromises always begin at *In the beginning God created* and, minus the COS†, the rest of the Bible becomes negotiable.

When the *Submerging Church* compromises the clear meaning of the Bible with gap theories, or any other old-earth, non-Scriptural teaching about origins, our youth, our peers and our society see the hypocrisy. And nowhere is this more apparent than in the workplace.

> **Mark 7:6-7**
> **6 This people honoureth me with their lips, but their heart is far from me...**
> **7 in vain do they worship me, teaching for doctrines the commandments of men.**

Please don't misunderstand me. I'm not saying that every Christian needs to work in a 501c3 non-profit ministry. What I am saying is that your work, whatever that may be, is your ministry.

Martin Luther wrote:

> **"The entire world is full of service to God, not only the churches but also the home, the kitchen, the cellar, the workshop, and the field of the townsfolk and farmers."**

The reason that so many people say they cannot tell a Christian from a non-Christian is because 95% of today's professing Christians hold to a secular worldview. (Barna Research)

So, in most instances, there truly is little difference.

The Parable of the Talents teaches us that God expects us to use the talents, gifts and abilities which He has given to us in serving His kingdom.

> **Matthew 25:14-30**
> **14 For the kingdom of heaven is as a man traveling into a far country, who called his own servants, and delivered unto them his goods.**

15 And unto one he gave five talents, to another two, and to another one; to every man according to his several ability; and straightway took his journey.

16 Then he that had received the five talents went and traded with the same, and made them other five talents.

17 And likewise he that had received two, he also gained other two.

18 But he that had received one went and digged in the earth, and hid his lord's money.

19 After a long time the lord of those servants cometh, and reckoneth with them.

20 And so he that had received five talents came and brought other five talents, saying, Lord, thou deliveredst unto me five talents: behold, I have gained beside them five talents more.

21 His lord said unto him, Well done, thou good and faithful servant: thou hast been faithful over a few things, I will make thee ruler over many things: enter thou into the joy of thy lord.

22 He also that had received two talents came and said, Lord, thou deliveredst unto me two talents: behold, I have gained two other talents beside them.

23 His lord said unto him, Well done, good and faithful servant; thou hast been faithful over a few things, I will make thee ruler over many things: enter thou into the joy of thy lord.

24 Then he which had received the one talent came and said, Lord, I knew thee that thou art an hard man, reaping where thou hast not sown, and gathering where thou hast not strawed:

25 And I was afraid, and went and hid thy talent in the earth: lo, there thou hast that is thine.

26 His lord answered and said unto him, Thou

**wicked and slothful servant, thou knewest that I
reap where I sowed not, and gather where I have
not strawed:
27 Thou oughtest therefore to have put my
money to the exchangers, and then at my coming
I should have received mine own with usury.
28 Take therefore the talent from him, and give
it unto him which hath ten talents.
29 For unto every one that hath shall be given,
and he shall have abundance: but from him that
hath not shall be taken away even that which he
hath.
30 And cast ye the unprofitable servant into
outer darkness: there shall be weeping and
gnashing of teeth.**

According to this, how we use our talents to serve the King is the true test of our faith. Twice in Scripture we are told that the earth shall be filled with the knowledge of God as the waters cover the seas (Habakkuk 2:14).

Our Christian calling, which begins in Genesis where we are told to take dominion, and continues through to the Great Commission (Matt 28) where we are instructed to make disciples of all nations as we take our own thoughts captive to the obedience of Christ (2 Cor 10:5).

Our calling requires us to thoughtfully develop and use our talents to build God's kingdom on earth. In this light we must examine ourselves to see if we are a good and faithful servant or a wicked and slothful one. Only then we can repent and make changes where needed.

This principle is clearly seen in the workplace.

I have several Christian friends who run their own businesses. Most have told me, "My worst customers are self-proclaiming Christians."

And the rest of the world sees such hypocritical Christians as well.

**The *Submerging Church* is getting harder
to tell apart from the secular world.**

Christians have always had to deal with things that are vulgar, lewd or coarse, but for the most part, as when I was growing up, we could simply steer clear of them. Today that is practically impossible.

I have heard it said that God speaks to us through our ears while Satan uses our eyes. However, the *Submerging Church* uses video images and music similar to that which is found in our secular culture.

And why not, if it makes people attend?

When the *Submerging Church* compromises the first five word's of Scripture, *In the beginning God created*, as we've said, the rest of God's Word becomes negotiable as well.

Today, instead of being a beacon of light, guiding people to the Lord of Creation, the *Submerging Church* has succumbed to the peer pressure of the secular world. Church leaders and marketing directors are trying to win back their flocks by basing the Church on the ever-shifting sands of man's ever-changing philosophies.

> **Matthew 7:26-27**
> **26 And every one that heareth these sayings of
> mine, and doeth them not, shall be likened unto a
> foolish man, which built his house upon the
> sand:**
> **27: And the rain descended, and the floods came,
> and the winds blew, and beat upon that house;
> and it fell: and great was the fall of it.**

And great will be the fall of such a house which loses 85% or more of its own children.

*The Submerging Church.*

The apostle Paul clearly saw the importance of holding to a Biblical worldview.

**2 Corinthians 10:5 Casting down imaginations, and every high thing that exalteth itself against the knowledge of God, and bringing into captivity every thought to the obedience of Christ.**

By hanging on to various *millions of years* imaginations and other compromised teachings about our origins, all of which exalt the word of man above the Word of God, the *Submerging Church* has made the Christian worldview and Christ Himself irrelevant to the world that Jesus created.

And irrelevant to the people that were made in His image. This includes the majority of Christian-raised young adults.

Clearly something must be done.

In the chapters that follow we will discuss some of the ways to deal with the perils and pitfalls of rebuilding the Church upon the rock of a Biblical view.

**Matthew 7:24-25**
**24 Therefore whosoever heareth these sayings of mine, and doeth them, I will liken him unto a wise man, which built his house upon a rock:**
**25 And the rain descended, and the floods came, and the winds blew, and beat upon that house; and it fell not: for it was founded upon a rock.**

When I speak, my wife Joanna runs a resource table for those who want to learn the information we share, or want to *Make It Their Ministry* to copy and give away our DVDs.

While at a church in Colorado a Sunday School teacher was looking over the dinosaur books for children. She explained that she had been teaching a unit on dinosaurs based on a young-earth, six-day Creation. Then she added, "But I'm not sure what to tell the kids about when dinosaurs went extinct."

"When do you believe they went extinct?" asked Joanna.

"Well," she said, "I'm not sure if it was 60 or 65 million years ago. What do you think?"

Joanna suggested, "I think you need to learn the facts that Russ teaches and then re-teach that class."

# EIGHT

# Relevance 101

The *Submerging Church* has accepted that the Bible teaches us some great moral and spiritual lessons while, at the same time, accepting that God's Word is unreliable in many areas which relate to the world we physically live in.

The result is that many Christians are completely baffled about how the Bible could really be given to us from an all-powerful Creator God—and whether or not the Bible's moral teachings are actually relevant when it comes to guiding their lives.

The failure of the *Submerging Church* to stand firm and defend the Word of God, beginning with the evidence that supports the first few chapters of the Book of Genesis, has left both Believers and seekers scratching their heads over what to believe.

**What you believe and why you believe it DOES MATTER!**

Our worldview is not some mere abstract notion. It is extremely functional, influencing the way we live our lives, day in and day out. It guides how we see and shape the world around us.

The Biblical worldview is based on the teachings of the Lord Jesus Christ. His teachings are founded upon the COSt.

This worldview is more consistent, more trustworthy and more logical than any other belief system when all the facts are laid out on the table.

It works because it's TRUE. It wins out over all the other competitors because it gives reliable answers to life's great questions:

**Where did I come from?**
**What is my purpose?**
**And what happens when I die?**

If we adopt a counterfeit worldview it can lead to consequences that are difficult to live with at best, and quite possibly an eternity of complete and utter despair.

On the other hand, when we live our lives in the reality of the Biblical worldview, our lives will become more purposeful and we will be able to remain on that narrow path, led by the knowledge of God in this life, and in the life to come in heaven.

I am not saying that once we adopt and live out the Biblical worldview our lives suddenly become perfect here on earth. We are still made of flesh and live in a sin-filled world.

What I am saying is that the Bible provides the accurate lens through which we can clearly see the Truth of God and the world we live in. Thus our lives will be consistent when they are based on the Truth and reality of God's Word.

The *Submerging Church* seems to understand that Scripture is designed to be the guiding light for our lives, yet blots out the proof with the black highlighter of compromise – and you can highlight every part of Scripture you object to.

A prevailing misconception is that there is the world of *religion* and then there is the world of *science*, and never should the two meet. This lie has allowed secularists to teach their beliefs of *millions of years leading to Darwinism* as if they were scientific truth. This concept is called NOMA—Non-Overlapping Magesteria.

NOMA holds that science, religion, and philosophy are the three great ways of knowing, but that they are completely independent and do not discuss the same issues; i.e., non-overlapping.

However, the truth is that real science and the Bible are never in conflict with each other. This is why I often tell people that *real science is a Christian's best friend.*

The Church has spent decades of time, billions of dollars and millions of man-hours evangelizing the world while stressing our personal relationship with Jesus Christ.

But this priority on personal relationship and personal relationship alone can also be a shortcoming because it prevents us from seeing God's plan for us beyond salvation.

Professing Christ and saying you're a Christian is more than discipleship and doctrine.

True Christianity is a way of seeing and understanding all of *real* life.

It is a worldview, built on the foundation of the first verse of the Bible:

**Genesis 1:1 In the beginning God created the heaven and the earth.**

Everything that exists was created by God, finds its purpose in Him and answers to Him. He created all of nature and the laws that keep it intact. God created our minds, our bodies, our souls and our spirits.

Note: Understanding the differences of the soul and the spirit can be confusing so allow me to briefly explain them.

Though the soul and the spirit are connected they are separate (Hebrews 4:12).

People have a soul. Though the Bible speaks of the soul in many contexts, in its most basic sense, it is the non-physical part of our life, it is who we are. The life of the soul is removed at the time of physical death (Genesis 35:18; Jeremiah 15:2).

People also have a spirit, but are we not spirits. Our spirit is what gives us the ability to have an intimate relationship with God. Whenever the word *spirit* is used, it refers to the non-physical part of us that connects with God, who Himself is spirit (John 4:24).

Only Believers are said to be spiritually alive (1 Corinthians 2:11; Hebrews 4:12; James 2:26), while non-believers are spiritually dead (Ephesians 2:1-5; Colossians 2:13).

God created us to learn and to use logic and discernment (not just to have discernment, but to use it). The truth of every topic we consider, from morals to finances to science, can be discerned through God's Word.

There is nothing outside the realm of God.
God is all Truth.
According to Jesus Christ Himself, He claims to be God, which is a key point that so many in the Church either ignore or overlook.

**John 14:6 I AM THE WAY, THE TRUTH and THE LIFE.**

Jesus is the agent, the Creator of the world in which:

**Col. 1:16-17**
**16 For by him were all things created, that are in heaven, and that are in earth, visible and invisible, whether they be thrones, or dominions, or principalities, or powers: all things were created by him, and for him:**
**17 And he is before all things, and by him all things consist.**

Jesus is the Creator of the born-again life; the author and finisher of our faith:

**Hebrews 12:2**
**Looking unto Jesus the author and finisher of our faith; who for the joy that was set before him endured the cross, despising the shame, and is set down at the right hand of the throne of God.**

Jesus will be the Creator of the new and glorified bodies:

**1 Corinthians 15:51-53**
**51 Behold, I shew you a mystery; We shall not all sleep, but we shall all be changed,**
**52 In a moment, in the twinkling of an eye, at the last trump: for the trumpet shall sound, and the dead shall be raised incorruptible, and we shall**

**be changed.**
**53 For this corruptible must put on**
**incorruption, and this mortal must put on**
**immortality.**

And Jesus will be the Creator of the new heavens and new earth:

**Revelation 21:1**
**And I saw a new heaven and a new earth, for the**
**first heaven and first earth were passed away;**
**and there was no more sea.**

When we truly take in this Truth, we can see that the Christian faith cannot be condensed into a worship formula, a high tech facility, a new church parking lot or a twenty-minute Sunday morning message.

It is an all-encompassing worldview.

Christianity is based on the Biblical worldview and can't be limited to just one section of our lives or even to a *saving* experience. It is the total framework by which we are to live and breathe. This Biblical worldview is the complete life-filter by which we are to exist.

In order to stay on the narrow path we MUST be able to defend God's Word. Deny God's Word (and that is what Biblical compromise really is) and you will be walking blindly along a path which has had many stumbling blocks laid down to cause you to veer even further onto the broad way that leads to our destruction.

**Matthew 7:13 Enter ye in at the strait gate: for**
**wide is the gate, and broad is the way, that**
**leadeth to destruction, and many there be which**
**go in thereat:**

**The Greatest World War**

As I wrote extensively in *The Theft of America's Heritage* we are involved in the greatest world war of all time. And the *Submerging Church* is waving the white flag of surrender.

Under the ruse of being *non-essential, too divisive* or (gulp) *unloving*, the *Submerging Church* has left many Christians anchored to a secular world. It is much like having a millstone hung around their necks while swimming in the depths of the sea.

The secular crowd does not care if you attend your church and talk about *church* things, just as long as you adhere to their worldview once you leave the church premises and return to the *real* world of the workplace, education, entertainment, geology, economics, government, psychology, biology (the list goes on and on).

Brock Lee, a fellow Creationist who helped edit this report, stated:

> **When I was in 5th grade, we were learning about evolution. One girl said that she was taught that God created everything, and the teacher told all of us that, "we could believe that on Sundays, but had to believe evolution on the weekdays." Also, in the movie *Expelled: No Intelligence Allowed*, P.Z. Myers states that atheists don't want to take religion away, only to make it a recreational thing like knitting. He is saying that religion should be like "oh, gee, that was fun, but it's time to go back to the real world."**

And the *Submerging Church* has been quite compliant to secular wishes.

And folks see the hypocrisy.

————

I was giving a Creation seminar at a secular university in a very nice performance hall that I had rented at my own expense. During the question-and-answer period a student stood up and asked, "How in the world did *they* allow you to speak about this stuff on a secular campus?"

I replied, "I'm a taxpayer and I can rent this facility the same as any other speaker can. Besides," I added, "Your professors teach their religious beliefs all day long and no one questions them."

The young man thought about it for a second and sat back down.

---

You should know that I actually admire many Secular Humanists who oppose me. They are well prepared in their apologetics and usually well organized in their religious movement. They have patiently and methodically moved into all aspects of our culture. Taking over educational, scientific and media establishments was simply a work of evil genius. They are well funded and their conversion rate is stupendous.

Furthermore, they have done an excellent job of intimidating, and thus *submerging*, the Christian Church.

They are often outspoken in their zeal to amass converts to their *religion*. These evangelizers leave no stone unturned in their quest to get the world to bow down to their anti-God beliefs of both *millions of years* and *Darwinism*.

I have to admire their zealousness while, at the same time ask, "Where has the zeal gone from the pulpits of Christian churches in the United States and elsewhere?"

Shouldn't Christian pastors and other church leaders be ardent preachers of and defenders of the Bible, word for word and cover to cover?

## The Great Commission

If our Savior mandated that we **go into all the world**, I must ask:

How did we manage to lose our Christian-based culture in our own nation?

In the USA our educational establishment, media, government and our entire culture were once all Christian-based.

I see churches sending people all around the world yet going out into our own backyard seems to be considered a *non-essential.*

In just two generations America has slid down the slippery slope from being a Christian-based nation to becoming a Secular Humanistic society. We have truly sunk into the abyss of Postmodernism.

Today, no one idea or worldview is considered to be true or objective. Now, all viewpoints, all lifestyles, all beliefs and experiences are considered equal—to be tolerated, well, except Christianity which is now frowned upon.

When all ideas are considered to be equal no idea is really worth one's total allegiance. In this climate the *Submerging Church* has flourished.

Christian institutions teaching secular opinions and interpretations that are in direct conflict with the Word of God leave budding Christian leaders confused and unable to stand for the Truth…many don't even know what the Truth is!

––––––––

One of my ministry's board members told me of the time his mother asked their pastor why he never had the church conduct a Bible study.

The pastor responded it was because he never learned how to do Bible studies where he attended seminary.

No, I am not making this up – the *Submerging Church.*

––––––––

I feel great compassion for a pastor or church leader who seeks outside help from Believers in explaining Creation, evolution and age-of-the-earth issues. In reality very few leaders understand the subject of these important teachings and the vast amount of data that supports the Biblical view in each of the areas. Bringing in an outside expert in these topics is wise.

However, one thing that truly confounds me is when I run into a person who is in a leadership position within a Christian church yet, out of vain pride, is indifferent to the overwhelming information supporting God's Word.

Having an *I don't care* attitude when 85% of our kids are leaving the Church is hard to comprehend, and surprisingly, such attitudes dominate the *Submerging Church.*

I submit the following example of what I'm talking about.

It took an associate pastor three years to get me in to speak at the church he worked for.

Why did it take so long?

Because another associate pastor on the same church staff believed in *millions of years.*

The OK was finally given for me to speak, but only on what the church considered an off night.

I presented *50 Facts versus Darwinism* on a Friday night.

The associate pastor had told me to expect 100 people, but 450 showed up. Everyone was very excited and we were already discussing a return visit for the following month when the old-earth associate pastor got promoted to senior pastor. He quickly *eliminated* the position my friend held and I will not be allowed to speak there again.

Putting his vain pride first, he accepts the 85% loss of his flock's children. I have to call that evil.

Jesus described such pride-filled religious leaders as whitewashed tombs that were full of dead men's bones.

**Matthew 23:27-28**
**27 Woe unto you, scribes and Pharisees, hypocrites! For ye are like unto whited sepulchers, which indeed appear beautiful outward, but are within full of dead men's bones, and of all uncleaness.**
**28 Even so ye also outwardly appear righteous unto men, but within ye are full of hypocrisy and iniquity.**

Oh, they looked like beautiful shrines on the outside but it was just a veneer, a shallow covering. They cared about the ways of God and the people only as far as they gained from doing so.

The pastor mentioned on the previous page may appear to be a great Christian leader on the surface, the senior pastor of a large and prosperous church. However, he is certainly nothing more than a white-washed tomb which is full of the bones of 85% of the Christian children entrusted to his shepherding.

Here's another example.

A college kid asked me if I would speak again at the university he attended—the same college where the professor led the course designed to mock our ministry. I told him to go for it, as long as he could find a campus ministry to sign off on it so it wouldn't cost me $700 (if a campus ministry signs up as the "host" it only costs me $150).

He said that would not be a problem.

I got this email from the student the next day:

**Subject: Speaking at the college**

**Hey Russ, I talked to the leader of XXXX Ministry at the school. She said she could not write off on any form because such a lecture would go against the belief of the ministries' sponsors that there is a possibility that one day to God could in fact be *billions of years*.**

Due to this ministry's compromise of God's clear Word hundreds of Christian kids, many of whom were in the process of writing the Church off as irrelevant, were proven to be correct.

This personally offends me at several levels. I always thought that the purpose of organized religion was to educate people about their religion instead of limiting exposure to control what people believe.

Not only are the majority of today's Christian leaders unaware of the science and data that support a recent, young-earth Creation and the Biblical worldview, most could not possibly care any less.

They are either blind or indifferent to how the Church's failure to stand up for the *earthly things* has made the Christian Church irrelevant to their own children and to society.

It is true that no one comes to God apart from faith. However, the Christian faith is not an irrational leap into a dark abyss. The Bible offers realistic propositions backed by logic and evidence.

In today's world Christians must seek answers and defend that which is true—the Bible. Scripture is very clear on this point.

> **1 Peter 3:15 But sanctify the Lord God in your hearts: and be ready always to give an answer to every man that asketh you a reason of the hope that is in you with meekness and fear.**

According to this verse, Christians are to be able to articulate their beliefs humbly, thoughtfully, reasonably and Biblically. We must especially stand at the ready to *give an answer* to Secular Humanist teachings because it is our failure to do so that has also allowed these teachings to infiltrate the Church.

This requires that Christians acquire a solid understanding of the conflicting worldviews. Only then will we realize why *In the beginning God created* is so foundational to the Christian faith.

Jesus also calls us to love the Lord our God with all our heart and with all our soul and with all our mind (Matthew 22.37). We are to take every thought captive to the obedience of Christ (2 Corinthians 10:5).

After all, about 80% of the branches of modern science were founded by Creation-believing scientists in order to study God's Creation. These were men of great intellect, with a vast knowledge of the structured world they lived in and of Scripture.

A large majority of the writers and signers of the Declaration of Independence and the U.S. Constitution were Creation-believing

Christians. These also were men of great intellect, with extensive knowledge of law, language and Scripture.

Most of the incredible art from the 17th and 18th centuries which now resides in the National Art Gallery in Washington, DC contain Christian themes.

Over and over again in literature, history, science and government we have seen brilliant contributions by Christians who loved God with their hearts, souls and minds. These were men and women of superior intellect, aptitude and quite often strong faith.

Believers are still called to *take every thought captive to the obedience of Christ.*

Today there are thousands of scientists, doctors, dentists, researchers, lawyers, architects, artists, writers, educators, and other career professionals, who see the world through a Biblical worldview.

The Biblical worldview can go toe-to-toe with the secular worldview anytime the true facts are reviewed, with one huge improvement:

Believers have absolute TRUTH on their side.

The *Submerging Church* has, quite unfortunately, given up huge territorial rights to the TRUTH. They have surrendered their God-given terrain because of one ever so subtle lie:

**The earth is *billions of years* old.**

> **Genesis 3:13 And the LORD God said unto the woman, What is this that thou hast done? And the woman said, The serpent beguiled me, and I did eat.**

It's hard to comprehend that something that seems so *un-earth-shattering*, so *subtle*, could so change the destiny of the Church, of society and of billions of peoples' eternal destinations. But it has and will continue to affect humanity until the day that Jesus returns or until the *Submerging Church* realizes the destructive nature of this lie and returns to preaching and teaching the Truth of God's uncompromised Word.

If you are wondering how these secular religious beliefs managed to take control of our educational and scientific establishments, not to mention the majority of Christian colleges and seminaries, the answer is simple:

This is not of man; it is of Satan.

> **Ephesians 6:12 For we wrestle not against flesh and blood, but against principalities, against powers, against rulers of the darkness of this world, against spiritual wickedness in high places.**

That might sound a bit intimidating. Fortunately God provides us with protection.

> **Ephesians 6:13-16**
> **13 Wherefore take unto you the whole armor of God, that ye may be able to withstand in the evil day, and having done all, to stand.**
> **14 Stand therefore, having your loins girt about with truth, and having on the breastplate of righteousness;**
> **15 And your feet shod with the preparations of the gospel of peace;**
> **16 Above all taking the shield of faith, wherewith ye shall be able to quench all the fiery darts of the wicked.**

We have discussed our shield already but God also gave us a helmet and one weapon: the Word of God.

> **Ephesians 6:17 And take the helmet of salvation, and the sword of the Spirit, *which is the word of God*.** (emphasis mine)

It is interesting that, in using the helmet in this comparison, salvation is presented as something that *surrounds one's head*. No matter how a wearer of the helmet turns, the helmet (from the wearer's perspective) surrounds all that he sees, much like a person's worldview.

How do we use that sword, the Bible, to defend our Biblical worldview?

Whenever I am debating a non-Christian about *millions of years* or Darwinism I discuss scientific facts. Whenever I am discussing the age of the earth with a Christian I first use the Sword, the Word of God.

Old-earth beliefs do not come from within God's Word, they come from secular interpretations of the earth's strata layers. The Word of God slices and dices old-earth beliefs when applied in correct context and I cover these in *371 Days That Scarred Our Planet*.

How can the *Submerging Church* return to being our beacon of light, providing wisdom and knowledge so we can see the world, and conduct ourselves—whether in matters of finances, work, entertainment, media, government, laws, relationships, moral issues and more—according to the Biblical worldview?

Stay tuned.

We will explore that in the next chapter.

Two years ago I was invited to do an interview on a Christian television network. The host was a woman who was the president of a pro-life ministry.

As we taped the interview, we talked about issues involving Darwinism and *millions of years* philosophies.

This woman stated that the number one reason young woman felt abortion was okay was the *science* that they believed backed evolution. She said that time after time when she talked with young women who were considering aborting their child, or as she counseled women that had had an abortion, the issue of evolution came up.

I told her, "We need to be dealing with the evil root of *millions of years leading to Darwinism* rather than just dealing with the abortion fruit alone."

# NINE

# Relevance 202

We have established that young men and women are leaving the *Submerging Church* in droves.

Even sadder, Christian compromises of God's Word have led them to believe that the Bible is not the authoritative Word of their supposed Creator, making Biblical teachings and the Church irrelevant to their lives.

Shockingly, recent studies reveal that young adults who grew up attending Sunday Schools are *less likely* to give the Church another chance later in their lives than are kids who never attended church!

Yes, you read that correctly.

The book *Already Gone* by Answers in Genesis founder Ken Ham and consumer behavior research analyst C. Britt Beemer came out while we were writing this fourth report in The GENESIS Heritage Report Series.

Statistics compiled by Beemer's firm, America's Research Group, confirm the huge disconnect taking place between our children and their church experience.

This report, *The Submerging Church*, is a logical follow-up to *Already Gone*. We outline in this chapter steps that need to be taken to stem the tide of defections from today's Church.

But first here's a summary of the alarming trends that are helping decimate the Church, affecting nearly all denominations and non-denominational congregations throughout the United States.

In the first scientific study of its kind, the *Beemer Report* reveals startling facts discovered through 20,000 phone calls and detailed surveys of a thousand 20- to 29-year-olds who used to attend evangelical churches on a regular basis, but have since left it behind.

The results are shocking:

- Those who faithfully attended Sunday School as a child are more likely to leave the church than those who did not.
- Those who regularly attended Sunday School are more likely to believe that the Bible is less true.
- Those who regularly attended Sunday School are actually more likely to defend the legality of abortion and gay marriage.
- Those who regularly attended Sunday School are actually more likely to defend premarital sex.

These trends must be dealt with now. We have already lost most of one generation.

We are losing 85% of our own kids and, of the 15% that stay in the Church, only a handful hold to a Biblical worldview. That's correct. Studies report that only 4% of born-again teens hold a Biblical view (about the same as for Christian adults by the way).

It is clear that something must be done. It's time to face the foundational facts so we can begin to fix the problem and in the following pages I will outline some ideas and concrete steps that can be taken to deal with problems inside of the *Submerging Church*.

My wife was the *Christian Education Director* at a large church for many years. Joanna worked very hard teaching *K through 12* about God's Word.

Both of us have taught Sunday School and served as youth leaders as well. Now we can see from the results that Christian education needs to take a much different approach.

So where do we begin?

### Train up a Child

As Proverbs 22:6 says:

**Train up a child in the way he should go; and when he is old he will not depart from it.**

We are no longer a Christian-based nation. Our children, from a VERY early age, are saturated by the philosophy of *millions of years* of time, placing death long before mankind arrived on the scene.

Public schools on all levels, museums and National Parks are all proselytizers of *millions of years leading to Darwinism* and such teachings infiltrate children's cartoons, movies and books. We've become a secular-based society existing in a secular-based world. Therein lies the challenge.

### Where to Begin?

- With young children, every moment is a teaching moment. Long before children can read or write they are absorbing ideas and information. As we guide them through their formative years we must constantly filter what they learn through a Biblical worldview. This starts with *In the beginning God created.*

- For young children, the natural world is a wondrous place. For parents it is a tremendous teaching tool. When I took my grandson to Grand Canyon, I explained to him why the river could not have carved the Canyon. We also talked about the huge water flow that broke through the Kaibab Upwarp and carved out the Canyon quickly. As we stood on the rim looking down, my eight-year-old grandson said, *"You're right, there's no way that river could have carved out this canyon."* Then we talked about the Global Flood and how that formed the earth's strata layers. That led us to Noah and the ark and how animals only *bring forth after their kind.* My point is to use every moment as a teaching moment.

• Christian-based resources for all ages need to address the *millions of years* and *Darwinism* issues. Though there are several good resources, most of today's Christian educational materials ignore the earthly things (John 3:12). This failure is utterly astounding as we leave the things from the world in which our kids live...dinosaurs, fossils, canyons, strata layers, birds, trees, etc., to be explained to them through a secular worldview! **Until this is rectified the Church and its Biblically-based teachings will continue to be seen as irrelevant in the eyes of our youth.** As I mentioned earlier, we offer a variety of good books and DVDs that help build a Biblical worldview. Reading to your children is a great way to teach them a love of reading. Watching Biblically-based DVDs with your kids lets them know that what they are viewing is relevant even to you. Both activities show your kids you love them and that both they, and what they are learning, are worth your time. Choose resources that reinforce the Biblical worldview. As a note: there are excellent children's books that have a Biblical worldview on history, law, finances, just about every topic. Work with your church's leaders to ensure your Sunday Schools are reinforcing the Biblical worldview in their classes—making the Word of God relate to the earthly things your child deals with every single day.

• I am constantly amazed at how teachable and eager to learn young children can be. More than entertainment, sports, money, toys or other materialistic things, your kids want your time and attention. If, as a parent you are thinking you don't have enough time or money to help your children battle for their minds and souls, you need to re-think your priorities. By cutting out those things we think are *necessities* and focusing in on the things that will build a Biblical worldview you can train up your children to trust you and see God's Word as being relevant to their lives.

A young mother attended one of our teachings and told Joanna that her son was studying dinosaurs in his 4th grade public school class. He was to write a report based on dinosaurs so she got a dinosaur book from our resource table that was based on a recent, young-earth Creation.

That week she went to her son's teacher and told her, "We are Christians and we don't believe in evolution or *millions of years*, but I do want my son to learn about dinosaurs. Could my son use this book to do his report?"

The teacher looked over the book and said, "Your son can definitely use this book."

- This incident is a perfect example of what can be done when Christians stand up for their faith, backed by good resources. This young mother was brave, yet humble. She showed her young son that what they believed was correct and worth standing up for. She approached the teacher in an honest and polite way which resulted in the teacher seeing dinosaurs in a whole new light. Later, when her son gave his oral report to his classmates, they also got to hear the Biblical view on both dinosaurs and the age of the earth. And her son got to learn that he could stand up for his belief. Definitely a win-win situation all around.

- As your children get older, make sure they have friends and acquaintances that hold a Biblical worldview. Peer acceptance and support are a huge encouragement for anyone, especially youth. Seek out other families who hold to Christian values. Do not shelter your child from the Secular Humanist teachings – instead **show them why secular teachings do not mesh with honest research** and teach them how to defend God's Word by making sure they understand the evidences that support a recent, young-earth Creation and the overall Biblical worldview. Our DVDs are great ways to learn this information at a popular, easy-to-understand level. There are many great Creation family camps, trips and conferences that offer an opportunity for the whole family to learn together. Our Biblically-based Grand Canyon Bus Tours are a great example. Again, this is money very well spent and can be something the whole family works toward.

A man called me to say that he couldn't afford to take his family of four on our Grand Canyon Bus Tour and wanted to know why it was so expensive. (At the time secular tours were $167; ours was $79).

I suggested he build his summer vacation around the tour and he said he couldn't do that as they had already booked a condo in San Diego for a week.

He had spent $2,000 on the condo. Together with the travel costs, food, a day at the zoo and Sea World he was spending $4,000 for a secular vacation but $316 for our life-changing Christian tour was *out of his price range*.

Christians need to re-think their priorities.

• One last thought on children and learning. Don't underestimate a child's ability to understand an issue. When I go into a church to give a Creation seminar, I encourage the church to keep as many children in the service as possible. The experience of seeing parents and other adults learning right alongside of them is a valuable tool to let our children know that the information is important. This can also lead to many great family discussions. Although bits and pieces of information may be *over their heads* they can and do learn from our teachings. I know this because the questions they ask afterwards reveal their attentiveness. I have learned the hard way that some of the most difficult questions come from young children.

A five-year-old raised his hand during the post-seminar Q&A session at a Church where I was speaking. I thought *Oh, what a cute little guy* as I asked him what his question was.

He asked, "Why can birds fly?" Well, I had never given it any thought. Most of the crowd politely stifled their giggles as I said, "Well, that's how God made them."

A five-year-old girl sitting next to the boy raised her hand and asked, "Why can bats fly?"

I asked her, "Are the two of you related?" as the place erupted in laughter.

---

The primary influence on a child's development is their parents. This is a Biblical principle.

**Deuteronomy 6: 6-9**
**6 And these words, which I command thee this day, shall be in thine heart.**
**7 And thou shalt teach them to thy children, and shalt talk of them when thou sittest in thine house, and when thou walkest by the way, and when thou liest down, and when thou risest up.**

**8 And thou shalt bind them for a sign upon thine hand, and they shall be as frontlets between thine eyes.**
**9 And thou shalt write them upon the posts of thy house, and on thy gates.**

And

**Ephesians 6:4**
**And ye fathers, provoke not your children to wrath: but bring them up in the nurture and admonition of the Lord.**

The church and its Sunday School teachers are meant to assist parents, not take their place. Grandparents, friends and other relatives are also influential in a child's development.

But when the Biblical worldview is absent from their home, children will absorb whatever worldview they are exposed to. And in this world, that will be the Secular-Humanist worldview based on *millions of years leading to Darwinism.* It has been said that nature hates a vacuum, and so when one is created by not educating our kids in the Bible, it will be filled with the most prevalent *fill* possible. In our society, that is evolutionary dogma.

### Train up a Parent

CLEARLY, if we want our children to learn, defend and live the Biblical worldview, then we as parents need to set such an example through our lives.

With 95% of Christian adults holding to a secular worldview, we each need to examine ourselves.

> • Do you hold a Biblical worldview? The easiest way to determine whether or not you hold to a Biblical worldview is to ask yourself a

question – if you believe one thing, and the Bible says another, which is correct? Which one needs to change? If the answer is that the Bible is correct, then that is the first step to holding a Biblical worldview. Once that is established, reading Scripture will do the rest. Do you know and understand Scripture? Can you explain and defend the Biblical position on the largest attack your kids will face on God's Word: *millions of years leading to Darwinism*? If not, there is an abundance of excellent material to help you learn and become more knowledgeable in these and other areas of Biblical apologetics and application. Know *what you believe* and *why you believe it* for your own edification and for the sake of your children.

• Is teaching your family the Biblical worldview a top priority in your life? If not, review what have been your priorities and see how they line up to what the Bible says about being a spouse and parent. God's plan for families is perfect. Christians need to know that plan and submit ourselves to following it.

• Examine the television shows your family watches (or shut off the TV). The same goes for movies and the Internet. We need to develop the attitude of the Psalmist as stated in **Psalms 101:3: I will set no wicked thing before mine eyes: I hate the work of them that turn aside; it shall not cleave to me.** There are educational and fun DVDs that support the Biblical worldview which you and your family can enjoy and learn from, if you are diligent.

- Encourage your children to find their *niche.* Some children learn visually and some by hands-on experience. Some children love to read, some love to hike, others love to garden. And yes, some love computers. Get to know your children's gifts and their favorite way to learn. Then use that to get them to enjoy learning to hold to the Biblical worldview.

- Know that YOU ARE NOT ALONE. There are thousands of families who are teaching, living and defending the Biblical worldview. Many excellent websites and resources are readily available to you.

- Find a Bible-believing, young-earth Creation-teaching church and support that God-honoring pastor. This is not a denominational issue as I do not know of a single group that does not have some leaders that have been fooled into believing in *millions of years.*

I cannot tell you of the hundreds of people who come up to me to let me know that their pastor is a Bible-believing, young-earth Creationist and that he would love to have us come to speak at their church. However, once they mention this to their pastor they quickly find out that he is a *millions of years* guy who does not want his flock learning the information we share.

**A Colorado pastor called me to say he was planning to get a group of pastors together to have me visit and share our teachings with all of the Church Body in their town.**

**He was very excited as the local high school was a staunch promoter of Darwinism and had misled many Christian students. However, this Bible-believing church leader was about to get a taste of the *Submerging Church.***

**He attended a city-wide pastors meeting and announced his intention to bring me in to refute *millions of years leading to Darwinism* and show that God's Word is trustworthy and authoritative for guiding our lives.**

**As he put it, "Almost every pastor stood up and walked out of the meeting! And the three that remained had no interest whatsoever in having you visit. I was shocked!"**

**All I could say was, "Welcome to the *Submerging Church*."**

When seeking a Bible-believing church, key questions to ask both the pastor and the leadership board include:

- How old do you think the earth is?
- What do you think about the global flood?
- How often do you have a Creation speaker present teachings in your church?
- Does your Sunday School curriculum teach Biblical apologetics to your children and youth?

If the church has a children's, youth or associate pastor, be sure to have the same discussion with them. This is most vital if your child will be under this person's tutelage when at church.

It is vitally important that you take these steps to help your family embrace, defend and protect a Biblical worldview. Parents are the front line for teaching their children and your church should be a center of wisdom, knowledge and learning that supports the parents' Biblical worldview.

This brings us to our next section.

### Train up a Teacher

Leading a Sunday School, children's church or youth group is a great opportunity to minister to young people's lives. This takes a lot of effort when done properly and can be very fruitful in terms of helping young minds grow in their knowledge of the Lord. Still, churches usually struggle to find teachers.

Recent studies showing past efforts in these areas of Christian education have been doing more harm than good can be taken two ways:

1] as cause to quit trying; or
2] as a call to adjust what we have been teaching.

As a Christian educator you could be a HUGE agent of change in the resources your church uses in its teaching programs. You could lead the charge to get relevant curriculum and speakers into your church and begin the campaign to train children and their parents in the evidence and facts (apologetics) they need to defend a Biblical worldview.

Warning:

The waters in which the Church is *Submerging* are often murky and full of sharks (disguised as sheep)!

Before you approach your church's leadership you need to mentally prepare yourself for the fact that you just may discover they don't believe in the Biblical Creation, especially the relatively young earth portion of God's Creation. Even if they claim that they do, you may find out that they will not want to teach and defend the Biblical position (*non-essential*).

When using the questions above be forewarned that you just might find you are afloat in a sea of *Submergers*. If so, you will have to make the decision to stay and challenge the status quo, or find a church that embraces the Biblical worldview, including a recent, young-earth Creation, and support them. Should you opt to stay, I strongly suggest you set a time limit as to how long you remain if real change is not observed.

Pray for the courage (something that we will address in the next chapter) to allow God to use you as His beacon of light. If you truly know that God has called you to teach, set your face like flint to teach the non-compromised Word of God—even if that has to take place in a different facility.

• As an educator you need to learn and prepare yourself to teach good apologetics. Invest in your learning as you will need a solid understanding of many topics. Encourage your church to send you and other teachers to Worldview Conferences and lectures on the science that backs Creation. Again, there are many excellent resources available for all ages.

• Realize that your job is two-fold. You must teach your students: 1] the Bible; and 2] the information that defends and relates God's Word to the world they live in. Your students need to be able to defend the Biblical worldview out in the *real* world. Your efforts will reap great rewards for you and your students, both here and for eternity.

• Do not teach *Bible stories*. This is part of the overall problem. Teach the Bible for what it is— the true history book of God's Creation. Teach the Bible as God's book of laws and morals. Teach the Bible as relevant for guiding our lives. Show kids how the Bible relates to the world they live in by using data, evidence, maps, charts, globes, fossils and more. As an example, instead of saying we are going to read the story of Abraham this morning, say, *Today we are going to study the history of Abraham and the facts about his life.* The Bible is not a book of fairy tales. It is the true history of the universe and the looking glass through which we can correctly view the entire world.

One last word on Sunday School and Christian education. IT'S NOT JUST FOR CHILDREN. As I have already mentioned, 49% of pastors and 96% of Christian adults hold a secular worldview.

If children learn great facts in their class and go home to parents who cannot reinforce that learning, most of the instruction will soon be disregarded.

Adult education classes and family courses need to cover the same type of information that the children's teachings cover so all members of your church can build a Biblical worldview and learn to defend their faith.

The whole church, from cradle to grave, needs to be working toward strengthening their Biblical worldview.

### Train up a Pastor

Youth and college pastors who stand firm on the non-compromised Word of God have my greatest respect. They are working with young men and women who are BOMBARDED on a daily basis with evolution, *millions of years*, MTV, peer pressure, drugs, alcohol, premarital sex, abortion and that uneasiness that comes with realizing they will soon have to make difficult decisions about school and careers.

Youth and college pastors are usually hired to *relate* to the youth. They are usually younger and *cooler* and can speak *youth-talk*.

Unfortunately, whether or not they are rated as being a *good* leader is based on the answer to one question:

"How many youth did you have at last night's meeting?"

This needs to change as it causes leaders to accept compromise positions in order to get as many people to attend their group as possible.

This pressures youth and college pastors to present lessons that are *seeker-friendly* while avoiding controversial subjects which are often relevant to the key issues their students face almost daily.

---

We have some close friends who are college pastors for a large *seeker-friendly* church. They wanted to spend their lesson times talking about issues that were relevant to young adults, like Creation vs. evolution, worldviews, and abortion.

After some of the discussions became rather lively, truly delving into the issues these young adults were facing, the senior pastor ordered them not to discuss these topics again.

I offered to come and teach about Creation, evolution and the age of the earth issues, the science that backs God's Word, but their pastor said these teachings were *non-essentials* and too heated to spend time discussing.

He then gave them a set of DVDs with a warm 12-minute message that was full of forgiveness and patience. It left everyone feeling good about themselves as they then ended the night with a nice song.

The group went from 25 in attendance to a group of three. Their meetings had become *irrelevant*, a waste of time.

### *Submerging.*

Instead of teaching irrelevant messages in hopes of not starting heated debates, we should be teaching college and youth the facts that back the Biblical Creation and refute the secular teachings about our origins. We should teach them how to humbly but firmly debate and defend the Biblical worldview. Besides, it's obvious and demonstrable that people prefer Truth which causes controversy over fluff that leads to blandness.

A youth or college pastor who learns these apologetics or invites in speakers or takes the youth to conferences and seminars where they can learn these apologetics, is the one who is truly making a lasting imprint on the lives of the teens.

### Senior Pastors

> **1 Thessalonians 2:13: For this cause also thank we God without ceasing, because, when ye received the word of God which ye heard of us, ye received it not as the word of men, but as it is**

**in truth, the word of God, which effectually worketh also in you that believe.**

Please repeat after me:

I, _____ , do solemnly promise before Jesus Christ my Lord and Savior, to preserve, protect and defend the Bible and the Biblical worldview, so help me God.

Forget denominational boundaries, man-made, purpose-driven and seeker-friendly formulas. Studies show they do not work when it comes to developing Bible-believing Christians.

Such man-made efforts only work from a worldly standpoint of building larger businesses – the Christian Industry.

**I know that this is not your goal or you wouldn't be reading this book.**

It doesn't matter if your church is mega, big, medium or small. It doesn't matter if your church is mostly seniors, families, or a blend of both.

If you were taught *millions of years* or Darwinism in a Christian college or seminary you need to realize that such teachings DO NOT come from God's Word.

Humble yourself to teach what the Bible clearly tells us and do what you set out to do as a pastor:

**LEAD YOUR FLOCK.**

Yes, you will likely run into an elder or tither who gets upset. Be prepared to firmly yet lovingly explain why they need to read and accept God's Word. Offer them great supporting resources as well.

Do not allow anyone to threaten or undermine your ministry. You are the one called to shepherd your flock and you will be the one held responsible for how you lead them. Sometimes being a leader means saying goodbye.

Your command is to preach, teach and defend the Word of God.

Word for word and cover to cover.

It is a demonstrable fact that you will lose more than 4 out of 5 of the kids in your congregation if you do not commit to making the Bible relevant to the world they live in.

> • Set the foundation of your church on the rock, the uncompromised Word of God. Know the facts that support Biblical Creation and all the other foundational issues (both historically and Scripturally) that are defined in the Book of Genesis.

> • Make sure all your staff members are completely on board with a recent, young-earth Creation. This is so important. Make sure your elders, deacons, trustees and teachers are completely versed in the Biblical worldview. If they are not, they need to become students in these areas and in the apologetics. If they refuse to drop *millions of years* philosophies, they should not be in a leadership position at your church.

Hey, no one said that being the leader was easy and fun. Go back and read about Moses. The Israelites nearly drove him crazy at times.

> • Every word of Scripture is relevant to your church members' lives. Your preaching must equip your flock with rock-solid Biblical Truths and the evidences needed for your flock to defend the Biblical worldview in their schools and workplaces.

> • If your church has been *Submerging*, you can change that. It will require some tough

decisions on your part. The first one is: are you willing to be that driving force for change? Are you willing to use your God-given authority to require that your staff members hold a Biblical worldview? Are you willing to require that your staff support you in this change and that they become learned in defending the Biblical worldview? These are tough choices to be sure, but they shouldn't be, as holding a Biblical worldview should be the only acceptable view for a Christian to hold. However, these choices are very necessary today if you want to win the battle for the hearts, minds and souls of those whom God has entrusted to you.

• Refine all that you do. Remember, you are in a war of worldviews and until you and your flock are completely up to battle-ready conditions, you should not waver or weave from preaching and teaching the Biblical worldview and how to defend it in our secular society.

• One word: Apologetics. The church needs to really step up to the plate on this. Classes, DVDs, conferences, speakers, seminars, books, tapes, resources and more are at your disposal. Leave no stone unturned. Whether it's your women's Bible Study or your youth group's outing, it should be centered on teaching participants to see things through a Biblical worldview while revealing the relevance of the Bible and Church in today's secular society. DO NOT EVER FORGET that your flock is constantly inundated with secular humanism, evolution, *millions of years* and Post-modernism on a continual basis. It never lets up; we are under a never-ending attack.

- Your family doctor calls in a specialist to deal with specific healthcare issues. Likewise, God has provided you with a world of God-honoring specialists to assist you in specific areas. This is especially true regarding apologetics. Do not hesitate to bring in outside speakers who specialize in particular areas. This is imperative in order to be an effective and God-honoring leader.

- Be prepared for some *Submergers* to jump ship. You might even have to do some budget reorganizing. Still, there is no substitute for straightforward Biblical teaching. In the end you will gain and develop true Believers who will be encouraged and strengthened in everything they do. And your leadership will make God's Word and your church relevant to the lives of those who attend your teachings; to those whom you shepherd.

Know that you are not alone. There are many pastors who are leading their church families to do exactly what is outlined above.

When I step into these churches and talk with the people there, you can feel God's Spirit at work and it is so uplifting.

To know God's Truth and to walk in that Truth takes a Christian to a whole new level of conviction, confidence and freedom.

This world is our God-given providence and territory.

Let's reclaim it!

Christian school leaders who stand firm for the Scriptural account of Creation, including the global flood judgment, really like our Grand Canyon Bus Tours. It is a really neat field trip for their students. One such group from Tucson, Arizona took a trip with us in late May 2008.

Well, low and behold, the morning of the trip we awoke to six inches of snow. Of course, the kids from Tucson thought it was great. And it was, except that it was a little cold.

We ended up giving our on-the-rim presentation right outside of a very busy gift shop and viewing point. As Russ was speaking, a good crowd of onlookers gathered and listened in as he pointed out what the Bible teaches about the flood and how the evidence for the Grand Canyon supports the Bible. Several people nodded their approval.

As he spoke a couple of teenaged girls asked me, "Does anyone ever get angry at Mr. Miller for what he is teaching?"

"Sure," I replied, "all the time, but we keep on teaching it because it's true."

Courage is not the absence of fear. Courage is feeling the fear and doing what is right anyway.

*Joanna Miller*

# TEN

# Courage

The Book of Genesis is the most amazing historical record in the world. In Genesis, Moses recorded the creation of the world, the Fall of man, the Curse, the promise of the coming Redeemer (born of a virgin), and so much more. Genesis also contains the historical record of the global deluge. Today we can see the ensuing geological upheaval it produced. Also recorded in Genesis...

**There were giants in the earth in the pre-flood days.**

> **Genesis 6: 4-5**
> **4 There were giants in the earth in those days;**
> ***and also after that*, when the sons of God came in**
> **into the daughters of men, and they bare**
> **children to them, and same became mighty men**
> **which were of old, men of renown.**
> **5 And God saw that the wickedness of man was**
> **great in the earth and that every imagination of**
> **the thoughts of his heart was only evil**
> **continually.** (emphasis mine)

God's global judgment via a worldwide flood destroyed everything and everyone, except for Noah and his family and the animals aboard the ark. God had provided one means of salvation from His coming judgment and anyone could have accepted God's salvation by walking up that narrow plank and entering the ark through that one open door. Only Noah and his family believed God's Word and they were saved from destruction. Those that perished in the deluge included the Nephilim. It is not within the scope of this report to do a complete study on the Nephilim.

Suffice to say they were men of incredible size and strength who instilled fear and a sense of powerlessness in the people of God.

It was not long after the global flood that the giants, or Nephilim, reared their evil heads again. We can read of them in Deuteronomy 2:10, Joshua 15:13 and Numbers 13:13.

Historically, Joshua and Caleb fought and prevailed against the giants for **the Lord thy God was with them, withersoever they goeth,** as we are told in Joshua 1:9.

But the giants did not end there. In 1 Samuel 17 we read of David the shepherd boy killing Goliath the giant, a descendent of the Anakim race of giants.

David slew the giant named Goliath after shouting these words:

> **1 Samuel 17:45 Thou comest to me with a sword, and with a spear, and with a shield: but I come to thee in the name of the Lord of hosts, the God of the armies of Israel, who thou hast defied.**

David killed this satanic giant because God was with him. God was guiding his life (and the rock coming from his sling).

**As in the Days of Noah**

> **Matthew 24: 37**
> **As in the days of Noah were, so shall also the coming of the Son of man be.**

As in the days of Noah, there are many giants today. Looming large are Secular-Humanist philosophies, secular school teachings and our secular media. Together they nurture and push a giant anti-Christian agenda.

There are also the giants of apathy, ignorance and indifference which reside, among other places, inside the *Submerging Church.*

And let's not forget the giants of materialism, over-indulgence, greed, violence, wickedness and immorality.

All of these giants have produced large amounts of evil fruit in our society.

Never has there been a world more in need of our Lord and Savior, Jesus Christ.

When we look at God's provisions we can see that He had provided a young boy named David with a leather sling and a stone.

God has provided today's Church Body with a powerful Sword and overwhelming evidences to the Truth of His Word.

The giants were about to defeat the ancient Israelites who were cowering in fear until a little shepherd boy named David stepped forth and defeated the giants of his day.

The *Submerging Church* is cowering in fear while being defeated by today's giants. The *Submerging Church* is surrendering to the enemy with scarcely a whimper.

This is because young David had something the *Submerging Church* sorely lacks:

**True faith in God and in His Word.**

Are we out of time to fight the giants that are prevailing against us? Is the Church too *Submerged* to be reclaimed?

Well, though it is late, it is not too late; not if we return to holding to a Biblical worldview while allowing God to guide our lives, our churches and other Christian institutions.

It is not too late if enough shepherds will pick up their Sword to take on the giants lined up against the Church today.

My prayer is that this book would be a charge for the Body of Christ to arm itself with our mighty and powerful Sword, the Word of God. My hope is that Scripture would encompass every step we take in our giant-filled world.

My prayer continues for the Church Body to make the Church relevant once again by only supporting Church leaders who trust, teach and defend all of God's Word, word for word and cover to cover, starting with **In the beginning God created.**

**Joshua 1:9**
**Have not I commanded thee? Be strong and of a good courage; be not afraid, neither be thou dismayed: for the Lord thy God is with thee whithersoever thou goest.**

---

## Make this your ministry

My hope is that many of you will review our life-changing teachings, which we've prepared in DVD format, and then...

### ...create your own DVD ministry.

Even if, like me, you can't chew gum and walk at the same time you can do this. It is one of the simplest, easiest, and most cost-effective ministries we know of.

And it gets results.

God can use each of us in this manner to reach many people and to reap a bountiful harvest of saved souls.

We've seen a number of people slough off their old-earth-belief shackles after being exposed to the truth presented in our DVDs.

Many of these people are professing Christians who had simply conformed to the false secular interpretations they'd been taught in school.

Others have become Believers after seeing our presentations and realizing—as I did when I got freed from Theistic Evolution—that the Bible really is true word for word and cover to cover.

All you have to do to begin your own ministry is select which DVD or DVD series you want to acquire, and then make copies and give them away for free.

That is correct; we want you to copy and give away our DVDs. In fact, we encourage you and others to soldier up with CESM and *Make This Your DVD Ministry.*

All of the details are on our web site:

www.creationministries.org

Go there and click on Create Your Own DVD Ministry.

If you're feeling a nudge that you should get into some type of ministry perhaps this is the ministry that nudge is directing you to consider.

Please go over the details on our web site and pray about making this your ministry.

If you read the info about me in **About the authors** near the front of this report you know that I gave away a lucrative business to found Creation, Evolution & Science Ministries.

And, yes, it is sometimes challenging but at the same time it is the most satisfying thing I've ever done.

Others who've soldiered up with us have told us the same thing.

Will you be the next one to share this experience?

**Matthew 9:37-38 Then saith he unto his disciples, The harvest truly is plenteous, but the labourers are few; Pray ye therefore the Lord of the harvest, that he will send forth labourers into his harvest.**

## The GENESIS Heritage Report Series

This series of book-length reports is a partnership effort between UCS PRESS and Creation, Evolution & Science Ministries. Go to www.new-earth-thought.com for ordering information and to register to receive for FREE our **50 Facts vs. Darwinism** e-mail series. Here are the titles of the trade paperback editions in The GENESIS Heritage Report Series:

### Report No. 1

## The Theft of America's Heritage

**Biblical foundations under siege:
a nation's freedoms vanishing**

### Report No. 2

## The Facts Are Talking, But Who's Listening?

**Refuting Darwinism – in Seven Seconds Flat!**

### Report No. 3

## 371 Days That Scarred Our Planet

**What the stones and bones reveal might surprise you.**

### Report No. 4

## The Submerging Church

**Eroded and made irrelevant by compromise**

Breinigsville, PA USA
20 November 2009
227933BV00004B/2/P